Aircraft Digital Electronic and Computer Systems:
Principles, Operation and Maintenance

Fly-by-wire: the Airbus A380 making its debut at the Farnborough International Airshow in July 2006

Aircraft Digital Electronic and Computer Systems:
Principles, Operation and Maintenance

Mike Tooley

AMSTERDAM • BOSTON • HEIDELBERG • LONDON • NEW YORK • OXFORD
PARIS • SAN · DIEGO • SAN FRANCISCO • SINGAPORE • SYDNEY • TOKYO

Butterworth-Heinemann is an imprint of Elsevier

ELSEVIER

Butterworth-Heinemann is an imprint of Elsevier
Linacre House, Jordan Hill, Oxford OX2 8DP, UK
30 Corporate Drive, Suite 400, Burlington, MA 01803, USA

First edition 2007

British Library Cataloguing in Publication Data
A catalogue record for this book is available from the British Library

Library of Congress Cataloging-in-Publication Data
A catalog record for this book is available from the Library of Congress

For information on all Elsevier Butterworth-Heinemann publications
visit our web site at books.elsevier.com

Typeset by the author

ISBN–13: 978-0-7506-8138-4
ISBN–10: 0-7506-8138-1

Interactive questions and answers, and other supporting material is available online.
To access this material please go to **www.key2study.com/66web** and follow the
instructions on screen

Contents

Preface

The books in this series have been designed for both independent and tutor assisted studies. They are particularly useful to the 'self-starter' and to those wishing to update or upgrade their aircraft maintenance licence. The series also provides a useful source of reference for those taking *ab initio* training programmes in EASA Part 147 and FAR 147 approved organizations as well as those following related programmes in further and higher education institutions.

This book is designed to cover the essential knowledge base required by certifying mechanics, technicians and engineers engaged in engineering maintenance activities on commercial aircraft. In addition, this book should appeal to members of the armed forces and others attending training and educational establishments engaged in aircraft maintenance and related aeronautical engineering programmes (including BTEC National and Higher National units as well as City and Guilds and NVQ courses).

The book provides an introduction to the principles, operation and maintenance of aircraft digital electronic and computer systems. The aim has been to make the subject material accessible and presented in a form that can be readily assimilated. The book provides full syllabus coverage of Module 5 of the EASA Part-66 syllabus with partial coverage of avionic topics in Modules 11 and 13. The book assumes a basic understanding of aircraft flight controls as well as an appreciation of electricity and electronics (broadly equivalent to Modules 3 and 4 of the EASA Part-66 syllabus).

Chapter 1 sets the scene by providing an overview of flight instruments and cockpit layouts. It also introduces the use of Electronic Flight Instruments (EFIS) and the displays that they produce.

Denary, binary and hexadecimal number systems are introduced in Chapter 2. This chapter provides numerous examples of the techniques used for converting from one number system to another, for example binary to hexadecimal or octal to binary.

Data conversion is the subject of Chapter 3. This chapter introduces analogue and digital signals and the techniques used for analogue to digital and digital to analogue conversion. Representative circuits are provided for various types of converter including successive approximation, flash and dual slope analogue to digital converters.

Chapter 4 describes the data bus systems that allow a wide variety of avionic equipment to communicate with one another and exchange data. The principles of aircraft bus systems and architecture are discussed and the operation of the ARINC 429 bus is discussed in detail. Various other bus standards (e.g. ARINC 629 and ARINC 573) are briefly discussed. Further references to aircraft bus systems (including those based on optical fibres) appear in later chapters.

Logic circuits are introduced in Chapter 5. This chapter begins by introducing the basic logic functions (AND, OR, NAND, and NOR) before moving on to provide an introduction to Boolean algebra and combinational logic arrangements. An example of the use of combinational logic is given in the form of a landing gear door warning system. Chapter 5 also describes the use of tri-state logic devices as well as monostable and bistable devices. An example of the use of combinational logic is included in the form an APU starter control circuit. The chapter concludes with an explanation of the properties and characteristics of common logic families, including major transistor-transistor logic (TTL) variants and complementary metal oxide semiconductor (CMOS) logic.

Modern aircraft use increasingly sophisticated avionic systems based on computers. Chapter 6 describes the basic elements used in a computer system and explains how data is represented and stored within a computer system. Various types of semiconductor memory are explained including random access memories (RAM) and read-only memories (ROM). The chapter also provides an introduction to computer programs and software and examples of computer instructions are given. Chapter 6 also provides an introduction to the backplane bus systems used for larger aircraft computers. The chapter is brought to a conclusion with a discussion of two examples of aircraft computers; a flight deck clock computer and an aircraft integrated data system (AIDS) data recorder.

Chapter 7 provides an introduction to the operation of microprocessor central processing units (CPU). The internal architecture of a typical CPU is presented together with a detailed explanation of its operation and the function of its internal elements.

Examples of several common microprocessor types are given, including the Intel x86 family, Intel Pentium family and the AMD 29050 which now forms the core of a proprietary Honeywell application specific integrated circuit (ASIC) specifically designed for critical embedded aerospace applications.

Chapter 8 describes the fabrication technology and application of a wide variety of modern integrated circuits, from those that use less than ten to those with many millions of active devices. The chapter also includes sections on the packaging and pin numbering of integrated circuit devices.

Medium Scale Integrated (MSI) logic circuits are frequently used in aircraft digital systems to satisfy the need for more complex logic functions such as those used for address decoding, code conversion, and the switching of logic signals between different bus systems. Chapter 9 describes typical MSI devices and their applications (including decoding, encoding and multiplexing). Examples of several common MSI TTL devices are included.

By virtue of their light weight, compact size, exceptional bandwidth and high immunity to electromagnetic interference, optical fibres are now widely used to interconnect aircraft computer systems. Chapter 10 provides an introduction to optical fibres and their increasing use in the local area networks (LAN) used in aircraft.

Chapter 11 describes typical displays used in avionic systems, including the cathode ray tube (CRT) and active matrix liquid crystal displays (AMLCD) used in Electronic Flight Instrument Systems (EFIS).

Modern microelectronic devices are particularly susceptible to damage from stray static charges and, as a consequence, they require special handling precautions. Chapter 12 deals with the techniques and correct practice for handling and transporting such devices.

Aircraft software is something that you can't see and you can't touch yet it must be treated with the same care and consideration as any other aircraft part. Chapter 13 describes the different classes of software used in an aircraft and explains the need for certification and periodic upgrading or modification. The chapter provides an example of the procedures required for upgrading the software used in an Electronic Engine Control (EEC).

One notable disadvantage of the increasing use of sophisticated electronics within an aircraft is the proliferation of sources of electromagnetic interference (EMI) and the urgent need to ensure that avionic systems are electromagnetically compatible with one another. Chapter 14 provides an introduction to electromagnetic compatibility (EMC) and provides examples of measures that can be taken to both reduce EMI and improve EMC. The chapter also discusses the need to ensure the electrical integrity of the aircraft structure and the techniques used for grounding and bonding which serves to protect an aircraft (and its occupants) from static and lightning discharge

Chapter 15 provides an introduction to a variety of different avionic systems based on digital electronics and computers. This chapter serves to bring into context the principles and theory discussed in the previous chapters.

The book concludes with three useful appendices, including a comprehensive list of abbreviations and acronyms used with aircraft digital electronics and computer systems.

The review questions at the end of each chapter are typical of those used in CAA and other examinations. Further examination practice can be gained from the four revision papers given in Appendix 2. Other features that will be particularly useful if you are an independent learner are the 'key points' and 'test your understanding' questions interspersed throughout the text.

Acknowledgements

The author would like to express sincere thanks to those who helped in producing this book. In particular, thanks go to Jonathan Simpson and Lyndsey Dixon from Elsevier who ably 'fielded' my many queries and who supported the book from its inception. Lloyd Dingle (who had the original idea for this series) for his vision and tireless enthusiasm. David Wyatt for proof reading the manuscript and for acting as a valuable 'sounding board'. Finally, a big 'thank you' to Yvonne for her patience, understanding and support during the many late nights and early mornings that went in to producing it!

Supporting material for this book (including interactive questions and answers) is available online. To access this material please go to **www.key2study.com/66web** and then follow the instructions on screen

Chapter 1 Introduction

Although it may not be apparent at first sight, it's fair to say that a modern aircraft simply could not fly without the electronic systems that provide the crew with a means of controlling the aircraft. Avionic systems are used in a wide variety of different applications ranging from flight control and instrumentation to navigation and communication. In fact, an aircraft that uses modern 'fly-by-wire' techniques could not even get off the ground without the electronic systems that make it work.

This chapter begins with an introduction to the basic instruments needed for indicating parameters such as heading, altitude and airspeed, and then continues by looking at their modern electronic equivalents. Finally, we show how flight information can be combined using integrated instrument systems and flight information displays, as shown in Figure 1.1.

1.1 Flight instruments

Of paramount importance in any aircraft is the system (or systems) used for sensing and indicating the aircraft's attitude, heading, altitude and speed. In early aircraft, these instruments were simple electro-mechanical devices. Indeed, when flying under Visual Flight Rules (VFR) rather than Instrument Flight Rules (IFR) the pilot's most important source of information about what the aircraft was doing would have been the view out of the cockpit window! Nowadays, sophisticated avionic and display technology, augmented by digital logic and computer systems, has made it possible for an aircraft to be flown (with a few possible exceptions) entirely by reference to instruments. More about these important topics appears in Chapters 5 and 6.

Various instruments are used to provide the pilot with flight-related information such as the aircraft's current heading, airspeed and attitude.

Figure 1.1 Boeing 757 flight instruments and displays

Modern aircraft use electronic transducers and electronic displays and indicators. Cathode ray tubes (CRT) and liquid crystal displays (LCD) are increasingly used to display this information in what has become known as a 'glass cockpit'. Modern passenger aircraft generally have a number of such displays including those used for primary flight data and multi-function displays that can be configured in order to display a variety of information. We shall begin this section with a brief review of the basic flight instruments.

1.1.1 Basic flight instruments

Crucial amongst the flight instruments fitted to any aircraft are those that indicate the position and attitude of the aircraft. These basic flight instruments are required to display information concerning:

- Heading
- Altitude
- Airspeed
- Rate of turn
- Rate of climb (or descent)
- Attitude (relative to the horizon).

A summary of the instruments that provide these indications is shown in Table 1.1 with the typical instrument displays shown in Figs. 1.2 to 1.8. Note that several of these instruments are driven from the aircraft's pitot-static system. Because of this, they are often referred to as 'air data instruments' (see Figs. 1.11 to 1.13).

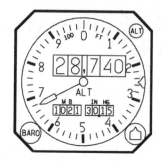

Figure 1.2   Altimeter

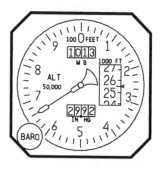

Figure 1.3   Standby altimeter

Table 1.1   Basic flight instruments

Instrument	Description
Altimeter (Fig. 1.2)	Indicates the aircraft's height (in feet or metres) above a reference level (usually mean sea level) by measuring the local air pressure. To provide accurate readings the instrument is adjustable for local barometric pressure. In large aircraft a second standby altimeter is often available (see Fig 1.3).
Attitude indicator or 'artificial horizon' (Fig. 1.4)	Displays the aircraft's attitude relative to the horizon (see Fig. 1.4). From this the pilot can tell whether the wings are level and if the aircraft nose is pointing above or below the horizon. This is a primary indicator for instrument flight and is also useful in conditions of poor visibility. Pilots are trained to use other instruments in combination should this instrument or its power fail.
Airspeed indicator (Figs. 1.5 and 1.6)	Displays the speed of the aircraft (in knots) relative to the surrounding air. The instrument compares the ram-air pressure in the aircraft's pitot tube with the static pressure (see Fig. 1.11). The indicated airspeed must be corrected for air density (which varies with altitude, temperature and humidity) and for wind conditions in order to obtain the speed over the ground.
Magnetic compass (Fig. 1.7)	Indicates the aircraft's heading relative to magnetic north. However, due to the inclination of the earth's magnetic field, the instrument can be unreliable when turning, climbing, descending, or accelerating. Because of this the HSI (see below) is used. For accurate navigation, it is necessary to correct the direction indicated in order to obtain the direction of true north or south (at the extreme ends of the earth's axis of rotation).
Horizontal situation indicator	The horizontal situation indicator (HSI) displays a plan view of the aircraft's position showing its heading. Information used by the HSI is derived from the compass and radio navigation equipment (VOR) which provides accurate bearings using ground stations. In light aircraft the VOR receiver is often combined with the VHF communication radio equipment but in larger aircraft a separate VOR receiver is fitted.
Turn and bank indicator or 'turn coordinator'	Indicates the direction and rate of turn. An internally mounted inclinometer displays the 'quality' of turn i.e. whether the turn is correctly coordinated, as opposed to an uncoordinated turn in which the aircraft would be in either a slip or skid. In modern aircraft the turn and bank indicator has been replaced by the turn coordinator which displays (a) rate and direction of roll when the aircraft is rolling, and (b) rate and direction of turn when the aircraft is not rolling.
Vertical speed indicator (Fig. 1.8)	Indicates rate of climb or descent (in feet per minute or metres per second) by sensing changes in air pressure (see Fig. 1.11).

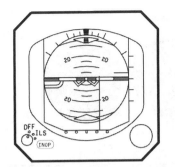

Figure 1.4   Attitude indicator

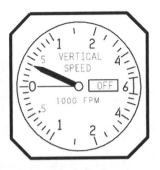

Figure 1.8   Standby vertical speed indicator

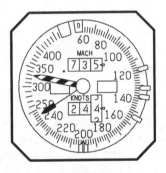

Figure 1.5   Airspeed indicator

Test your understanding 1.1

Identify the instruments shown in Figs.1.9 and 1.10. Also state the current indication displayed by each instrument.

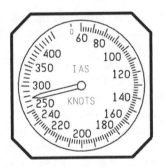

Figure 1.6   Standby airspeed indicator

Figure 1.9 See Test your understanding 1.1

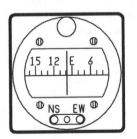

Figure 1.7   Standby magnetic compass

Figure 1.10 See Test your understanding 1.1

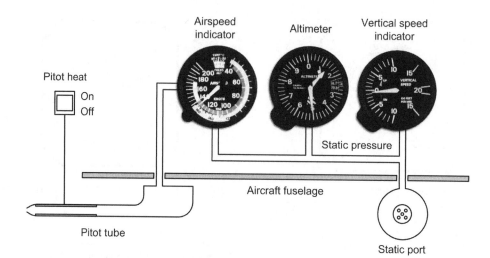

Figure 1.11 Pitot-static driven instruments

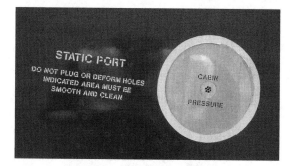

Figure 1.12 An aircraft static port

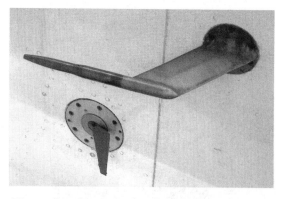

Figure 1.13 A pitot tube (upper right) and an angle of attack sensor (lower left)

1.1.2 Acronyms

A number of acronyms are used to refer to flight instruments and cockpit indicating systems. Unfortunately, there is also some variation in the acronyms used by different aircraft manufacturers. The most commonly used acronyms are listed in Table 1.2. A full list can be found in Appendix 1.

Table 1.2 Some commonly used acronyms

Acronym	Meaning
ADI	Attitude direction indicator
ASI	Airspeed indicator
CDU	Control and display unit
EADI	Electronic attitude and direction indicator
ECAM	Electronic centralized aircraft monitoring
EFIS	Electronic flight instrument system
EHSI	Electronic horizontal situation indicator
EICAS	Engine indicating and crew alerting system
FDS	Flight director system
FIS	Flight instrument system
FMC	Flight management computer
FMS	Flight management system
HSI	Horizontal situation indicator
IRS	Inertial reference system
ND	Navigation display
PFD	Primary flight display
RCDI	Rate of climb/descent indicator
RMI	Radio magnetic indicator
VOR	Very high frequency omni-range
VSI	Vertical speed indicator

1.1.3 Electronic flight instruments

Modern aircraft make extensive use of electronic instruments and displays. One advantage of using electronic instruments is that data can easily be exchanged between different instrument systems and used as a basis for automatic flight control. We will explore the potential of this a little later in this chapter but, for now, we will look at the two arguably most important electronic instruments, the electronic attitude and direction indicator (EADI) and the electronic horizontal situation indicator (EHSI).

Electronic attitude and direction indicator

The electronic attitude direction indicator (EADI—see Fig. 1.14) is designed to replace the basic ADI and normally comprises:

- an attitude indicator
- a fixed aircraft symbol
- pitch and bank command bars
- a glide slope indicator
- a localizer deviation indicator
- a slip indicator
- flight mode anunciator
- various warning flags.

The aircraft's attitude relative to the horizon is indicated by the fixed aircraft symbol and the flight command bars. The pilot can adjust the symbol to one of three flight modes. To fly the aircraft with the command bars armed, the pilot simply inserts the aircraft symbol between the command bars.

The command bars move up for a climb or down for descent, roll left or right to provide lateral guidance. They display the computed angle of bank for standard-rate turns to enable the

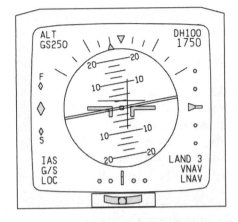

Figure 1.14   A typical EADI display

pilot to reach and fly a selected heading or track. The bars also show pitch commands that allow the pilot to capture and fly an ILS glide slope, a pre-selected pitch attitude, or maintain a selected barometric altitude. To comply with the directions indicated by the command bars, the pilot manoeuvres the aircraft to align the fixed symbol with the command bars. When not using the bars, the pilot can move them out of view.

The glide slope deviation pointer represents the centre of the instrument landing system (ILS) glide slope and displays vertical deviation of the aircraft from the glide slope centre. The glide slope scale centreline shows aircraft position in relation to the glide slope.

The localizer deviation pointer, a symbolic runway, represents the centre of the ILS localizer, and comes into view when the pilot has acquired the glide slope. The expanded scale movement shows lateral deviation from the localizer and is approximately twice as sensitive as the lateral deviation bar in the horizontal situation indicator. The selected flight mode is displayed in the lower left of the EADI for pitch modes, and lower right for lateral modes. The slip indicator provides an indication of slip or skid indications.

Electronic horizontal situation indicator

The electronic horizontal situation indicator (EHSI) assists pilots with the interpretation of information provided by a number of different navigations aids. There are various types of EHSI but essentially they all perform the same function. An EHSI display (see Fig. 1.15) can be configured to display a variety of information (combined in various different ways) including:

- heading indication
- radio magnetic indication (RMI)
- track indication
- range indication
- wind speed and direction
- VOR, DME, ILS or ADF information.

1.1.4 Flight director systems

The major components of a flight director system (FDS) are the electronic attitude and direction indicator (EADI) and electronic horizontal situation indicator (EHSI) working together with a mode selector and a flight director computer.

The FDS combines the outputs of the electronic flight instruments to provide an easily interpreted display of the aircraft's flight path. By comparing this information with the pre-programmed flight path, the system can automatically compute the necessary flight control commands to obtain and hold the desired path.

The flight director system receives information from the:

- attitude gyro
- VOR/localizer/glide slope receiver
- radar altimeter
- compass system
- barometric sensors.

The flight director computer uses this data to provide flight control command information that enables the aircraft to:

- fly a selected heading
- fly a predetermined pitch attitude
- maintain altitude
- intercept a selected VOR track and maintain that track
- fly an ILS glide slope/localizer.

The flight director control panel comprises a mode selector switch and control panel that provides the input information used by the FDS.

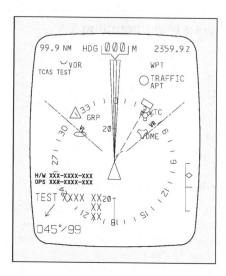

Figure 1.15   A typical EHSI display

The pitch command control pre-sets the desired pitch angle of the aircraft for climb or descent. The command bars on the FDS then display the computed attitude to maintain the pre-selected pitch angle. The pilot may choose from among many modes including the HDG (heading) mode, the VOR/LOC (localizer tracking) mode, or the AUTO APP or G/S (automatic capture and tracking of ILS and glide path) mode. The auto mode has a fully automatic pitch selection computer that takes into account aircraft performance and wind conditions, and operates once the pilot has reached the ILS glide slope.

Flight director systems have become increasingly more sophisticated in recent years. More information appears in Chapter 15.

Test your understanding 1.2

1. What are the advantages of Flight Director Systems (FDS)?

2. List FOUR inputs used by a basic FDS.

3. List FOUR types of flight control information that can be produced by a basic FDS.

4. Explain the function of the FDS auto mode during aircraft approach and landing.

5. Explain the use of the symbolic runway in relation to the display produced by the EADI.

1.1.5 Electronic flight instrument systems (EFIS)

An electronic flight instrument system (EFIS) is a system of graphically presented displays with underlying sensors, electronic circuitry and software that effectively replaces all mechanical flight instruments and gauges with a single unit.

The EFIS fitted to larger aircraft consists of a primary flight display (PFD) or electronic attitude and direction indicator (EADI) and a navigation display (ND) or electronic horizontal situation indicator (EHSI). These instruments are duplicated for the captain and the first officer.

The PFD presents the usual attitude indicator in connection with other data, such as airspeed, altitude, vertical speed, heading or coupled landing systems (see Fig. 1.16). The ND displays route information, a compass card or the weather radar picture (see Fig. 1.17).

In addition to the two large graphical displays, a typical EFIS will have a display select panel, a display processor unit, a weather radar panel, a multifunction processor unit, and a multifunction display. We will look briefly at each of these.

EFIS primary flight display (PFD)

The typical EFIS PFD is a multicolour cathode ray tube (CRT) or liquid crystal display (LCD) display unit that presents a display of aircraft attitude and flight control system commands including VOR, localizer, TACAN (Tactical Air Navigation), or RNAV (Area Navigation) deviation together with glide slope or pre-selected altitude deviation. Various other information can be displayed including mode annunciation, radar altitude, decision height and excessive ILS deviation.

EFIS navigation display (ND)

Like the EFIS PFD, a typical EFIS ND takes the form of a multicolour CRT or LCD display unit. However, in this case the display shows the aircraft's horizontal situation information which, according to the display mode selected, can include compass heading, selected heading, selected VOR, localizer, or RNAV course and deviation (including annunciation or deviation type), navigation source annunciation, digital

Figure 1.16 EFIS primary flight display

Figure 1.17 EFIS navigation display

selected course/desired track readout, excessive ILS deviation, to/from information, distance to station/waypoint, glide slope, or VNAV deviation, ground speed, time-to-go, elapsed time or wind, course information and source annunciation from a second navigation source, weather radar target alert, waypoint alert when RNAV is the navigation source, and a bearing pointer that can be driven by VOR, RNAV or

ADF sources as selected on the display select panel. The display mode can also be set to *approach format* or *en-route format* with or without weather radar information included in the display.

Display select panel (DSP)

The display select panel provides navigation sensor selection, bearing pointer selection, format selection, navigation data selection (ground speed, time-to-go, time, and wind direction/speed), and the selection of VNAV (if the aircraft has this system), weather, or second navigation source on the ND. A DH SET control that allows decision height to be set on the PFD is also provided. Additionally, course, course direct to, and heading are selected from the DSP.

Display processor unit (DPU)

The display processor unit provides sensor input processing and switching, the necessary deflection and video signals, and power for the electronic flight displays. The DPU is capable of driving two electronic flight displays with different deflection and video signals. For example, a PFD on one display and an ND on the other.

Weather radar panel (WXP)

The weather radar panel provides MODE control (OFF, STBY, TEST, NORM, WX, and MAP), RANGE selection (10, 25, 50, 100, 200 and 300 nm), and system operating controls for the display of weather radar information on the MFD and the ND when RDR is selected on the MFD and/or the DSP.

Multifunction Display (MFD)

The multifunction display takes the form of another multicolour CRT or active-matrix LCD display unit. The display is normally mounted on the instrument panel in the space provided for the weather radar (WXR) indicator. Standard functions displayed by the unit include weather radar, pictorial navigation map, and in some

systems, check list and other operating data. Additionally, the MFD can display flight data or navigation data in case of a PFD or ND failure.

Multifunction processor unit (MPU)

The multifunction processor unit provides sensor input processing and switching and the necessary deflection and video signals for the multifunction display. The MPU can provide the deflection and video signals to the PFD and ND displays in the event of failures in either or both display processor units.

1.1.6 Electronic centralized aircraft monitor (ECAM)

Technical information concerning the state of an Airbus aircraft is displayed using the aircraft's electronic centralized aircraft monitor (ECAM—see Fig. 1.18). This normally takes the form of two CRT or LCD displays that are vertically arranged in the centre of the instrument panel. The upper (primary) display shows the primary engine parameters (N1/fan speed, EGT, N2/high pressure turbine speed), as well as the fuel flow, the status of lift augmentation devices (flap and slat positions), along with other information. The lower (secondary) ECAM display presents additional information including that relating to any system malfunction and its consequences.

Figure 1.18 A320 ECAM displays located above the centre console between the captain and first officer

Figure 1.19 Boeing 757 EICAS display

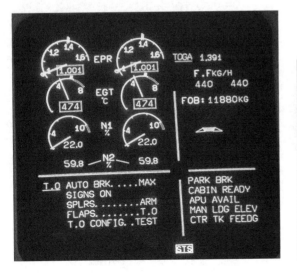

Figure 1.20 See Test your understanding 1.3

Test your understanding 1.3

Figure 1.20 shows a flight deck display.

1. Identify the display.

2. What information is currently displayed?

3. Where is the display usually found?

4. What fan speed is indicated?

5. What temperature is indicated?

1.1.7 Engine indicating and crew alerting system (EICAS)

In Boeing aircraft the equivalent integrated electronic aircraft monitoring system is known as the engine indicating and crew alerting system (EICAS). This system provides graphical monitoring of the engines of later Boeing aircraft, replacing a large number of individual panel-mounted instruments. In common with the Airbus ECAM system, EICAS uses two vertically mounted centrally located displays (see Fig. 1.19).

The upper (primary) EICAS display shows the engine parameters and alert messages whilst the lower (secondary) display provides supplementary data (including advisory and warning information). We shall be looking at the ECAM and EICAS systems in greater detail later in Chapter 15.

Figure 1.21 Captain's FMS CDU

1.1.8 Flight management system (FMS)

The flight management system (FMS) fitted to a modern passenger aircraft brings together data and information gathered from the electronic flight instruments, aircraft monitoring and navigation systems, and provides outputs that can be used for automatic control of the aircraft from immediately after take-off to final approach and landing. The key elements of an FMS include a flight management computer (FMC), control and display unit (CDU), IRS, AFCS, and a system of data buses that facilitates the interchange of data with the other digital and computerized systems and instruments fitted to the aircraft.

Two FMS are fitted, one for the captain and one for the first officer. During normal operation the two systems share the incoming data. However, each system can be made to operate independently in the event of failure. By automatically comparing (on a continuous basis) the indications and outputs provided by the two systems it is possible to detect faults within the system and avoid erroneous indications. The inputs to the FMC are derived from several other systems including IRS, EICAS, engine thrust management computer, and the air data computer. Figures 1.21 and 1.22 shows the FMC control and display units fitted to an A320 aircraft. We shall be looking at the operation of the FMS in greater detail later in Chapter 15.

1.2 Cockpit layouts

Major developments in display technology and the introduction of increasingly sophisticated aircraft computer systems have meant that cockpit layouts have been subject to continuous change over the past few decades. At the same time, aircraft designers have had to respond to the need to ensure that the flight crew are not over-burdened with information and that relevant data is presented in an appropriate form and at the time that it is needed.

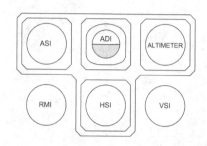

(a) Basic 'T' flight instrument configuration

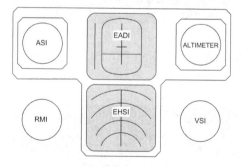

(b) Basic EFIS flight instrument configuration

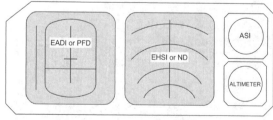

(c) Enhanced EFIS flight instrument configuration

Figure 1.22 A320 cockpit layout

Figure 1.23 Evolution of instrument layouts

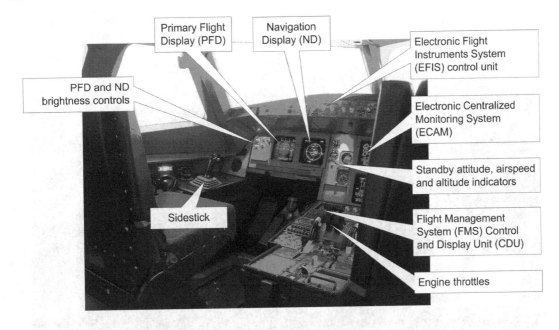

Figure 1.24 Captain's flight instrument and display layout on the A320

Figure 1.23 shows how the modern EFIS layouts have evolved progressively from the basic-T instrument configuration found in non-EFIS aircraft. Maintaining the relative position of the instruments has been important in allowing pilots to adapt from one aircraft type to another. At the same time, the large size of modern CRT and LCD displays, coupled with the ability of these instruments to display combined data (for example, heading, airspeed and altitude) has led to a less-cluttered instrument panel (see Fig. 1.24). Lastly, a number of standby (or secondary) instruments are made available in order to provide the flight crew with reference information which may become invaluable in the case of a malfunction in the computer system.

1.3 Multiple choice questions

1. A multi-function display (MFD) can be:
   (a) used only for basic flight information
   (b) configured for more than one type of information
   (c) set to display information from the standby magnetic compass.

2. Which standby instruments are driven from the aircraft's pitot-static system?
   (a) airspeed indicator, altimeter, and vertical speed indicator
   (b) airspeed indicator, vertical speed indicator, and magnetic compass
   (c) airspeed indicator, altimeter, and angle of attack indicator.

3. Static pressure is fed:
   (a) only to the airspeed indicator
   (b) only to the airspeed indicator and vertical speed indicator
   (c) to the airspeed indicator as well as the altimeter and vertical speed indicator.

4. The horizontal situation indicator (HSI) uses information derived from:
   (a) the VOR receiver
   (b) the pitot-static system
   (c) the airspeed indicator.

5. The term 'glass cockpit' refers to:
   (a) the use of LCD and CRT displays
   (b) the use of toughened glass windows
   (c) the use of a transparent partition between the flight deck and passenger cabin.

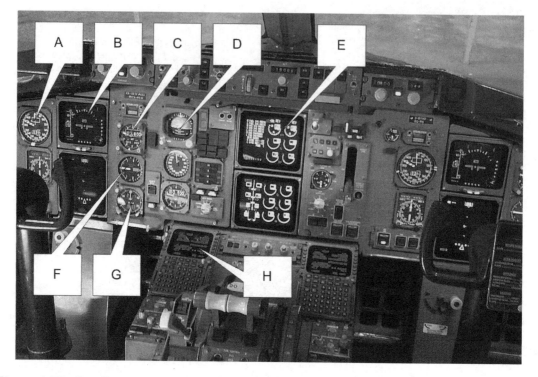

Figure 1.25 See Test your understanding 1.4

Test your understanding 1.4

1. Identify each of the Boeing 767 flight instruments and displays shown in Fig. 1.25.

2. Classify the flight instruments in Question 1 as either primary or standby.

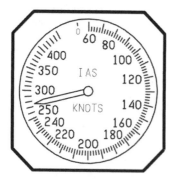

Figure 1.26 See Question 8

6. Basic air data instruments are:
   (a) airspeed indicator, altimeter, and magnetic compass
   (b) airspeed indicator, altimeter, and vertical speed indicator
   (c) airspeed indicator, vertical speed indicator, and artificial horizon.

7. Three airborne parameters that can be used to assess aircraft position are:
   (a) airspeed, height and weight
   (b) heading, airspeed and height
   (c) heading, weight and airspeed.

8. The instrument shown in Fig. 1.26 is the:
   (a) ADI
   (b) ASI
   (c) VSI

9. Engine parameters such as turbine speed, are displayed on:
   (a) ECAM
   (b) EHSI
   (c) EADI.

10. The aircraft slip indicator is found in the:
    (a) EADI
    (b) ECAM
    (c) CRTs in the passenger cabin.

11. Where is the 'rising runway'?
    (a) EADI
    (b) ECAM
    (c) CRTs in the passenger cabin.

12. Under visual flight rules (VFR) the pilot's most important source of information concerning the aircraft's position and attitude is:
    (a) the view out of the cockpit window
    (b) the altimeter and vertical speed indicators
    (c) the airspeed indicator and the magnetic compass.

13. The EFIS fitted to a large aircraft usually consists of:
    (a) a single multi-function display
    (b) separate primary flight and navigation displays
    (c) a primary display with several standby instruments.

14. The instrument shown in Fig. 1.27 is:
    (a) EADI
    (b) EHSI
    (c) ECAM.

Figure 1.28 See Question 15

15. The instrument shown in Fig. 1.28 is:
    (a) EADI
    (b) EHSI
    (c) ECAM.

16. In a basic T-configuration of instruments:
    (a) the ADI appears on the left and the ASI appears on the right
    (b) the ADI appears on the left and the HSI appears on the right.
    (c) the ASI appears on the left and the altimeter appears on the right.

17. The flight director system receives information:
    (a) only from the VOR/localizer
    (b) only from the attitude gyro and altimeter
    (c) from both of the above.

18. EICAS provides the following:
    (a) engine parameters only
    (b) engine parameters and system warnings
    (c) engine parameters and navigational data.

19. The two sets of flight regulations that a pilot may fly by are:
    (a) VFR and IFR
    (b) VHF and IFR
    (c) VFR and IFU.

20. Secondary heading information is obtained from:
    (a) the gyro
    (b) the compass
    (c) the pitot-static system.

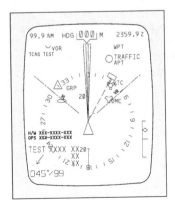

Figure 1.27 See Question 14

21. A major advantage of EFIS is a reduction in:
    (a) effects of EMI
    (b) wiring and cabling
    (c) moving parts present in the flight deck.

22. The term 'artificial horizon' is sometimes used to describe the indication produced by:
    (a) the altimeter
    (b) the attitude indicator.
    (c) the vertical speed indicator.

23. Typical displays on an EHSI are:
    (a) engine indications
    (b) VOR, heading, track
    (c) VOR, altitude, rate of climb.

24. In a basic T-configuration of instruments:
    (a) the ADI appears at the top and the HSI appears at the bottom
    (b) the HSI appears at the top and the ASI appears at the bottom
    (c) the ASI appears at the top and the ADI appears at the bottom.

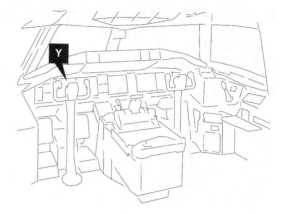

Figure 1.30   See Question 27

25. Operational faults in FMS can be detected by:
    (a) automatically comparing outputs on a continuous basis
    (b) routine maintenance inspection of the aircraft
    (c) pre-flight checks.

26. The display marked X in Fig. 1.29 is the:
    (a) navigation display
    (b) primary flight display
    (c) FMS CDU.

27. The display marked Y in Fig. 1.30 is the:
    (a) standby flight instruments
    (b) primary flight display
    (c) FMS CDU.

28. The upper ECAM display provides:
    (a) navigation information
    (b) secondary flight information
    (c) engine parameters.

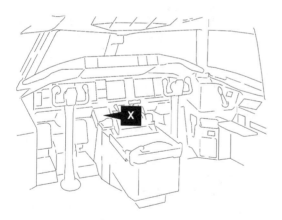

Figure 1.29   See Question 26

Chapter 2 — Number systems

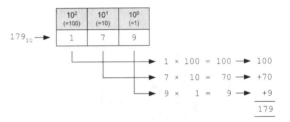

Figure 2.1 An example showing how the decimal number 179 is constructed

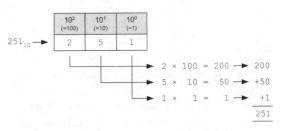

Figure 2.2 An example showing how the decimal number 251 is constructed

The signals in digital logic and computer systems are conveyed along individual electrical conductors and also using multiple wiring arrangements where several conductors are used to convey signals from one place to another in what is known as a **bus system**. As we will see in Chapter 4, the number of individual bus lines depends upon the particular bus standard employed however signals on the individual lines, no matter what they are used for nor how they are organised, can exist in only two basic states: logic 0 ('low' or 'off') or logic 1 ('high' or 'on'). Thus information within a digital system is represented in the form of a sequence of 1s and 0s known as **binary data**.

Since binary numbers (particularly large ones) are not very convenient for human use, we often convert binary numbers to other forms of number that are easier to recognise and manipulate. These number systems include **hexadecimal** (base 16) and **octal** (base 8). This chapter is designed to introduce you to the different types of number system as well as the process of conversion from one type to another.

Test your understanding 2.1

Write down the values of:
(a) 2×10^2
(b) 3×10^4
(c) $(1 \times 10^3) + (9 \times 10^2) + (0 \times 10^1) + (1 \times 10^0)$.

2.1 Decimal (denary) numbers

The decimal numbers that we are all very familiar with use the base 10. In this system the **weight** of each digit is 10 times as great as the digit immediately to its right. The rightmost digit of a decimal **integer** (i.e. a whole number) is the unit's place (10^0), the digit to its left is the ten's digit (10^1), the next is the hundred's digit (10^2), and so on. The valid digits in a decimal number are 0 to 9. Figures 2.1 and 2.2 show two examples of how decimal numbers are constructed. Note that we have used the suffix '10' to indicate that the number is a decimal. So, 179_{10} and 251_{10} are both decimal (or base 10) numbers. The use of subscripts helps us to avoid confusion about what number base we are actually dealing with.

2.2 Binary numbers

In the binary system (base 2), the weight of each digit is two times as great as the digit immediately to its right. The rightmost digit of a binary integer is the one's digit, the next digit to the left is the two's digit, next is the four's digit, then the eight's digit, and so on. The valid digits in the binary system are 0 and 1. Figure 2.3 shows an example of a binary number (note the use of the suffix '2' to indicate the number base).

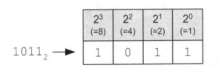

Figure 2.3 Example of a binary number.

Table 2.1 Binary and decimal numbers

Binary	Dec.
0000	0
0001	1
0010	2
0011	3
0100	4
0101	5
0110	6
0111	7
1000	8
1001	9

The binary numbers that are equivalent to the decimal numbers 0 to 9 are shown in Table 2.1. Notice how the **most significant digit (MSD)** is shown on the left and the **least significant digit (LSD)** appears on the right. In the table, the MSD has a weight of 2^3 (or 8 in decimal) whilst the LSD has a weight of 2^0 (or 1 in decimal). Since the MSD and LSD are represented by binary digits (either 0 or 1) we often refer to them as the **most significant bit (MSB)** and **least significant bit (LSB)** respectively, as shown in Fig. 2.4.

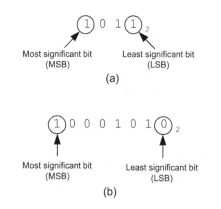

Figure 2.4 MSB and LSB in binary numbers

Test your understanding 2.2

1. What is the binary value of (a) the MSB and (b) the LSB in the binary number 101100?
2. What is the binary weight of the MSB in the number 10001101?

2.2.1 Binary to decimal conversion

In order to convert a binary number to its equivalent decimal number we can determine the value of each successive binary digit, multiply it by the column value (in terms of the power of the base) and then simply add the values up. For example, to convert the binary number 1011, we take each digit and multiply it by the binary weight of the digit position (8, 4, 2 and 1) and add the result, as shown in Fig. 2.5.

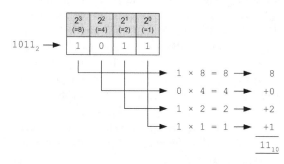

Figure 2.4 Example of binary to decimal conversion

2.2.2 Decimal to binary conversion

There are two basic methods for converting decimal numbers to their equivalent in binary. The first method involves breaking the number down into a succession of numbers that are each powers of 2 and then placing the relevant digit (either a 0 or a 1) in the respective digit position, as shown in Fig. 2.6.

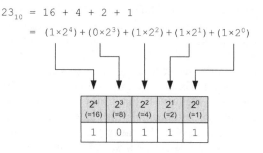

Figure 2.6 Example of decimal to binary conversion

Another method involves successive division by two, retaining the remainder as a binary digit and then using the result as the next number to be divided, as shown in Figure 2.7. Note how the binary number is built up in reverse order i.e. with the last remainder as the MSB and the first remainder as the LSB.

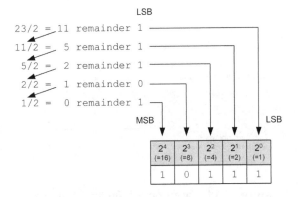

Figure 2.7 Example of decimal to binary conversion using successive division

Test your understanding 2.3

1. Convert the following binary numbers to decimal:
   (a) 10101
   (b) 110011
   (c) 1001001
   (d) 10101011.
2. Convert the following decimal numbers to binary:
   (a) 25
   (b) 43
   (c) 65
   (d) 100.

2.1.3 Binary coded decimal

The system of binary numbers that we have looked at so far is more correctly known as **natural binary**. Another form of binary number commonly used in digital logic circuits is known as **binary coded decimal (BCD)**. In this simpler system, binary conversion to and from decimal numbers involves arranging binary numbers in

groups of four binary digits from right to left, each of which corresponds to a single decimal digit, as shown in Figures 2.8 and 2.9.

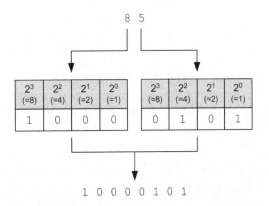

Figure 2.8 Example of converting the decimal number 85 to binary coded decimal (BCD)

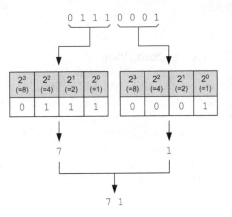

Figure 2.9 Example of converting the BCD number 01110001 to decimal

Test your understanding 2.4

1. Convert the following decimal numbers to binary coded decimal (BCD):
   (a) 97
   (b) 6450
2. Convert the following binary coded decimal (BCD) numbers to decimal:
   (a) 10000011
   (b) 011111101001

2.2.4 One's complement

The one's complement of a binary number is formed by **inverting** the value of each digit of the original binary number (i.e. replacing 1s with 0s and 0s with 1s) So, for example, the one's complement of the binary number 1010 is simply 0101. Similarly, the one's complement of 01110001 is 10001110. Note that if you add the one's complement of a number to the original number the result will be all 1s, as shown in Figure 2.10.

Original binary number:	1 0 1 1 0 1 0 1
One's complement:	+ 0 1 0 0 1 0 1 0
Added together:	1 1 1 1 1 1 1 1

Figure 2.10 The result of adding the one's complement of a number to the original number

2.2.5 Two's complement

Two's complement notation is frequently used to represent negative numbers in computer mathematics (with only one possible code for zero—unlike one's complement notation). The two's complement of a binary number is formed by inverting the digits of the original binary number and then adding 1 to the result. So, for example, the two's complement of the binary number 1001 is 0111. Similarly, the two's complement of 01110001 is 10001111. When two's complement notation is used to represent negative numbers the most significant digit (MSD) is always a 1. Figure 2.11 shows two examples of finding the two's complement of a binary number. In the case of Figure 2.11(b) it is

important to note the use of a carry digit when performing the binary addition.

Original binary number:	1 0 1 1 0 1 0 1
One's complement:	+ 0 1 0 0 1 0 1 0
Adding 1:	+ 0 0 0 0 0 0 0 1
Two's complement:	0 1 0 0 1 0 1 1

(a)

Original binary number:	1 0 0 1 1 1 0 0
One's complement:	+ 0 1 1 0 0 0 1 1
Adding 1:	+ 0 0 0 0 0 0 0 1
	1 1 carry
Two's complement:	0 1 1 0 0 1 0 0

(b)

Figure 2.11 Method of finding the two's complement of a binary number

2.3 Octal numbers

The octal number system is used as a more compact way of representing binary numbers. Because octal consists of eight digits (0 to 7), a single octal digit can replace three binary digits. Putting this another way, by arranging a binary number into groups of three binary digits (or **bits**) we can replace each group by a single octal digit, see Figure 2.12. Note that, in a similar manner to the numbering systems that we met previously in this chapter, the weight of each digit in an octal number is eight times as great as the digit immediately to its right. The rightmost digit of an octal number is the unit's place (8^0), the digit to its left is the eight's digit (8^1), the next is the 64's digit (8^2), and so on.

Test your understanding 2.5

1. Find the one's complement of the binary number 100010.
2. Find the two's complement of the binary number 101101.

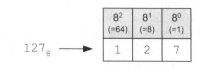

Figure 2.11 Example of an octal number

2.3.1 Octal to decimal conversion

In order to convert a binary number to a decimal number we can determine the value of each successive octal digit, multiply it by the column value (in terms of the power of the base) and simply add the values up. For example, the octal number 207 is converted by taking each digit and then multiplying it by the octal weight of the digit position and adding the result, as shown in Figure 2.13.

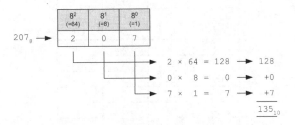

Figure 2.13 Example of octal to decimal conversion

2.3.2 Decimal to octal conversion

As with decimal to binary conversion, there are two methods for converting decimal numbers to octal. The first method involves breaking the number down into a succession of numbers that are each powers of 8 and then placing the relevant digit (having a value between 0 and 7) in the respective digit position, as shown in Figure 2.14.

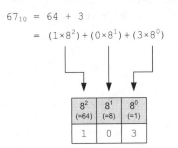

Figure 2.14 Example of decimal to octal conversion

The other method of decimal to octal conversion involves successive division by eight, retaining

the remainder as a digit (with a value between 0 and 7) before using the result as the next number to be divided, as shown in Figure 2.15. Note how the octal number is built up in reverse order i.e. with the last remainder as the MSD and the first remainder as the LSD.

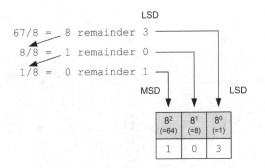

Figure 2.15 Example of decimal to octal conversion using successive division

2.3.3 Octal to binary conversion

In order to convert an octal number to a binary number we simply convert each digit of the octal number to its corresponding three-bit binary value, as shown in Fig. 2.16.

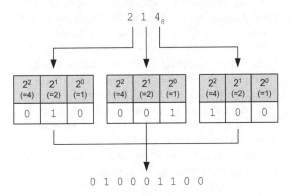

Figure 2.16 Example of octal to binary conversion

2.3.4 Binary to octal conversion

Converting a binary number to its equivalent in octal is also extremely easy. In this case you simply need to arrange the binary number in groups of three binary digits from right to left and then convert each group to its equivalent octal number, as shown Fig. 2.17.

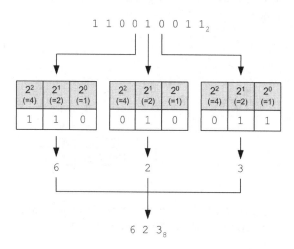

Figure 2.17   Example of binary to octal conversion

Table 2.2 Binary, decimal, hexadecimal and octal numbers

Binary	Dec.	Hex.	Octal
0000	0	0	0
0001	1	1	1
0010	2	2	2
0011	3	3	3
0100	4	4	4
0101	5	5	5
0110	6	6	6
0111	7	7	7
1000	8	8	10
1001	9	9	11
1010	10	A	12
1011	11	B	13
1100	12	C	14
1101	13	D	15
1110	14	E	16
1111	15	F	17

2.4 Hexadecimal numbers

Although computers are quite comfortable working with binary numbers of 8, 16, or even 32 binary digits, humans find it inconvenient to work with so many digits at a time. The hexadecimal (base 16) numbering system offers a practical compromise acceptable to both to humans and to machines. One hexadecimal digit can represent four binary digits, thus an 8-bit binary number can be expressed using two hexadecimal digits. For example, 10000011 binary is the same as 83 when expressed in hexadecimal.

The correspondence between a hexadecimal (hex) digit and the four binary digits it represents is quite straightforward and easy to learn (see Table 2.2). Note that, in hexadecimal, the decimal numbers from 10 to 15 are represented by the letters A to F respectively. Furthermore, conversion between binary and hexadecimal is fairly straightforward by simply arranging the binary digits in groups of four bits (starting from the least significant). Hexadecimal notation is much more compact than binary notation and easier to work with than decimal notation.

2.4.1 Hexadecimal to decimal conversion

In order to convert a hexadecimal number to a decimal number we can determine the value of each successive hexadecimal digit, multiply it by the column value (in terms of the power of the base) and simply add the values up. For example, the hexadecimal number of A7 is converted by taking each digit and then multiplying it by the weight of the digit position, as shown in Figure 2.18.

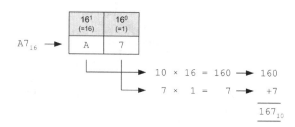

Figure 2.18   Example of hexadecimal to decimal conversion

2.4.2 Decimal to hexadecimal to conversion

In order to convert a decimal number to its hexadecimal equivalent you can break the number down into a succession of numbers that are each powers of 16 and then place the relevant digit (a value between 0 and F) in the respective digit position, as shown in Figure 2.19. Note how, in the case of the example shown in Figure 2.19(b) the letters F and E respectively replace the decimal numbers 15 and 14.

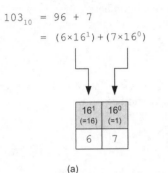

$$103_{10} = 96 + 7$$
$$= (6 \times 16^1) + (7 \times 16^0)$$

(a)

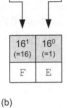

$$254_{10} = 240 + 14$$
$$= (15 \times 16^1) + (14 \times 16^0)$$

(b)

Figure 2.19 Example of decimal to hexadecimal conversion

2.4.3 Hexadecimal to binary conversion

In order to convert a hexadecimal number to a binary number we simply need to convert each digit of the hexadecimal number to its corresponding four-bit binary value, as shown in Figure 2.20. The method is similar to that which you used earlier to convert octal numbers to their

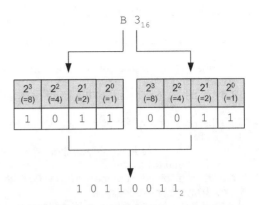

Figure 2.20 Example of hexadecimal to binary conversion

2.4.4 Binary to hexadecimal conversion

Converting a binary number to its equivalent in hexadecimal is also extremely easy. In this case you simply need to arrange the binary number in groups of four binary digits working from right to left before converting each group to its hexadecimal equivalent, as shown Figure 2.21. Once again, the method is similar to that which you used earlier to convert binary numbers to their octal equivalents.

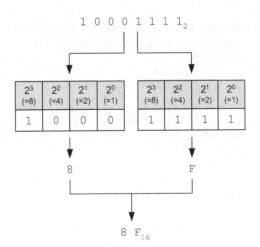

Figure 2.21 Example of binary to hexadecimal conversion

Test your understanding 2.6

1. Find the decimal equivalent of the octal number 41.
2. Find the octal equivalent of the decimal number 139.
3. Find the binary equivalent of the octal number 537.
4. Find the octal equivalent of the binary number 111001100.
5. Convert the hexadecimal number 3F to (a) decimal and (b) binary.
6. Convert the binary number 101111001 to (a) octal and (b) hexadecimal.
7. Which of the following numbers is the largest? (a) $C5_{16}$ (b) 11000001_2 (c) 303_8.

2.5 Multiple choice questions

1. The binary number 10101 is equivalent to the decimal number:
   (a) 19
   (b) 21
   (c) 35.

2. The decimal number 29 is equivalent to the binary number:
   (a) 10111
   (b) 11011
   (c) 11101.

3. Which one of the following gives the two's complement of the binary number 10110?
   (a) 01010
   (b) 01001
   (c) 10001.

4. The binary coded decimal (BCD) number 10010001 is equivalent to the decimal number:
   (a) 19
   (b) 91
   (c) 145.

5. The decimal number 37 is equivalent to the binary coded decimal (BCD) number:
   (a) 00110111
   (b) 00100101
   (c) 00101111.

6. Which one of the following numbers could NOT be an octal number?
   (a) 11011
   (b) 771
   (c) 139.

7. The octal number 73 is equivalent to the decimal number:
   (a) 47
   (b) 59
   (c) 111.

8. The binary number 100010001 is equivalent to the octal number:
   (a) 111
   (b) 273
   (c) 421.

9. The hexadecimal number 111 is equivalent to the octal number:
   (a) 73
   (b) 273
   (c) 421.

10. The hexadecimal number C9 is equivalent to the decimal number:
    (a) 21
    (b) 129
    (c) 201.

11. The binary number 10110011 is equivalent to the hexadecimal number:
    (a) 93
    (b) B3
    (c) 113.

12. The hexadecimal number AD is equivalent to the binary number:
    (a) 10101101
    (b) 11011010
    (c) 10001101.

13. The number 706_8 is equivalent to:
    (a) $1C6_{16}$
    (b) 111001110_2
    (c) 484_{10}.

14. The number 101110000_8 is equivalent to:
    (a) 160_{16}
    (b) 570_8
    (c) 368_{10}.

Chapter 3   Data conversion

Because signals in the real world exist in both digital (on/off) and analogue (continuously variable) forms, digital and computer systems need to be able to accept and generate both types of signal as inputs and outputs respectively. Because of this, there is a need for devices that can convert signals in analogue form to their equivalent in digital form, and vice versa. This chapter introduces digital to analogue and analogue to digital conversion. Later we will show how data conversion devices are used in some practical aircraft systems. We shall begin by looking at the essential characteristics of analogue and digital signals and the principle of **quantization**.

3.1 Analogue and digital signals

Examples of analogue and digital signals are shown in Figure 3.1. The analogue signal shown in Figure 3.1(a) consists of a continuously changing voltage level whereas the digital signal shown in Figure 3.1(b) consists of a series of discrete voltage levels that alternate between logic 0 ('low' or 'off') and logic 1 ('high' or 'on'). The actual voltages used to represent the logic levels are determined by the type of logic circuitry employed however logic 1 is often represented by a voltage of approximately +5V and logic 0 by a voltage of 0V (we will discuss the actual range of voltages used to represent logic levels later in Chapter 5).

In order to represent an analogue signal using digital codes it is necessary to approximate (or **quantize**) the signal into a set of discrete voltage levels as shown in Figure 3.2.

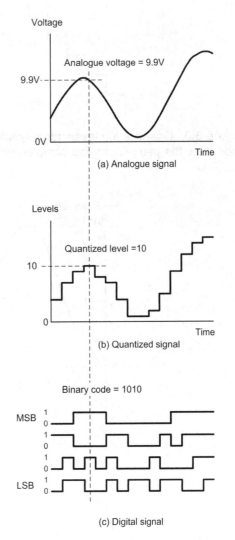

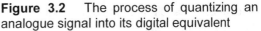

Figure 3.2   The process of quantizing an analogue signal into its digital equivalent

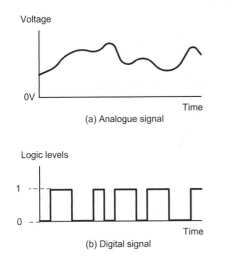

Figure 3.1   Example of (a) analogue and (b) digital signals

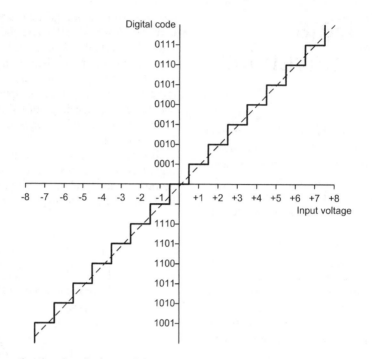

Figure 3.3 Quantization levels for a simple analogue to digital converter using a four-bit binary code (note the use of the two's complement to indicate negative voltage levels)

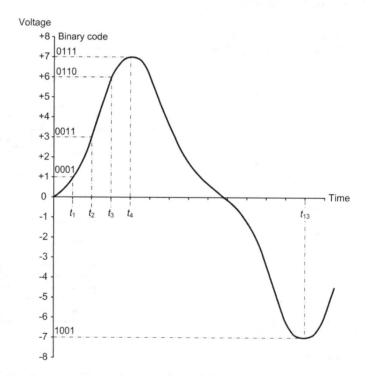

Figure 3.4   A bipolar analogue signal quantized into voltage levels by sampling at regular intervals (t_1, t_2, t_3, etc.)

The sixteen quantization levels for a simple analogue to digital converter using a four-bit binary code are shown in Fig. 3.3. Note that, in order to accommodate analogue signals that have both positive and negative polarity we have used the two's complement representation to indicate negative voltage levels. Thus, any voltage represented by a digital code in which the MSB is logic 1 will be negative. Figure 3.4 shows how a typical analogue signal would be quantized into voltage levels by **sampling** at regular intervals (t_1, t_2, t_3, etc).

3.2 Digital to analogue conversion

The basic digital to analogue converter (DAC) has a number of digital inputs (often 8, 10, 12, or 16) and a single analogue output, as shown in Figure 3.5.

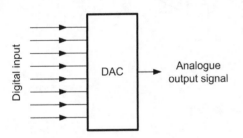

Figure 3.5   Basic DAC arrangement

The simplest form of digital to converter shown in Fig. 3.6(a) uses a set of binary weighted resistors to define the voltage gain of an operational summing amplifier and a four-bit binary latch to store the binary input whilst it is being converted. Note that, since the amplifier is connected in inverting mode, the analogue output voltage will be negative rather than positive. However, a further inverting amplifier stage can be added at the output in order to change the polarity if required.

The voltage gain of the inputs to the operational amplifier (determined by the ratio of feedback to input resistance and taking into account the inverting configuration) is shown in Table 3.1. If we assume that the logic levels produced by the four-bit data latch are 'ideal'

Table 3.1   Voltage gain for the simple DAC shown in Figure 3.6(a)

Bit	Voltage gain
3 (MSB)	$-R/R = -1$
2	$-R/2R = -0.5$
1	$-R/4R = -0.25$
0 (LSB)	$-R/8R = -0.125$

Table 3.2   Output voltages produced by the simple DAC shown in Fig.3.6(a)

Bit 3	Bit 2	Bit 1	Bit 0	Output voltage
0	0	0	0	0V
0	0	0	1	−0.625V
0	0	1	0	−1.250V
0	0	1	1	−1.875V
0	1	0	0	−2.500V
0	1	0	1	−3.125V
0	1	1	0	−3.750V
0	1	1	1	−4.375V
1	0	0	0	−5.000V
1	0	0	1	−5.625V
1	0	1	0	−6.250V
1	0	1	1	−6.875V
1	1	0	0	−7.500V
1	1	0	1	−8.125V
1	1	1	0	−8.750V
1	1	1	1	−9.375V

(such that logic 1 corresponds to +5V and logic 0 corresponds to 0V) we can determine the output voltage corresponding to the eight possible input states by summing the voltages that will result from each of the four inputs taken independently. For example, when the output of the latch takes the binary value 1010 the output voltage can be calculated from the relationship:

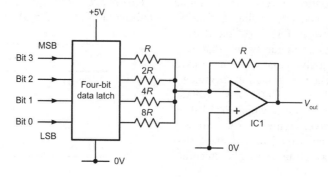

(a) Simple binary-weighted DAC

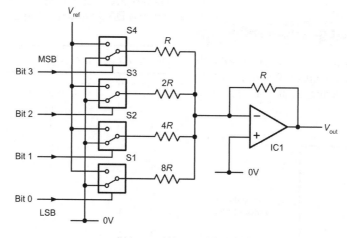

(b) Improved binary-weighted DAC

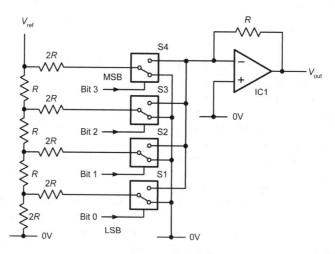

(c) R-2R ladder DAC

Figure 3.6 Simple DAC arrangements

$$V_{out} = \quad (-1 \times 5) + (-0.5 \times 0) + (-0.25 \times 5)$$
$$+ (-0.125 \times 0) = -6.25V$$

Similarly, when the output of the latch takes the binary value 1111 (the maximum possible) the output voltage can be determined from:

$$V_{out} = \quad (-1 \times 5) + (-0.5 \times 5) + (-0.25 \times 5)$$
$$+ (-0.125 \times 5) = -9.375V$$

The complete set of voltages corresponding to all eight possible binary codes are given in the Table 3.2.

An improved binary-weighted DAC is shown in Fig. 3.6(b). This circuit operates on a similar principle to that shown in Fig. 3.6(a) but uses four analogue switches instead of a four-bit data latch. The analogue switches are controlled by the logic inputs so that the respective output is connected to the reference voltage (V_{ref}) when the respective logic input is at logic 1 and to 0V when the corresponding logic input is at logic 0. When compared with the previous arrangement, this circuit offers the advantage that the reference voltage is considerably more accurate and stable than using the logic level to define the analogue output voltage. A further advantage arises from the fact that the reference voltage can be made negative in which case the analogue output voltage will become positive. Typical reference voltages are –5V, –10V, +5V and +10V.

Unfortunately, by virtue of the range of resistance values required, the binary weighted DAC becomes increasingly impractical for higher resolution applications. Taking a 10-bit circuit as an example, and assuming that the basic value of R is 1 kΩ, the binary weighted values would become:

Bit 0   512 kΩ
Bit 1   256 kΩ
Bit 2   128 kΩ
Bit 3   64 kΩ
Bit 4   32 kΩ
Bit 5   16 kΩ
Bit 6   8 kΩ
Bit 7   4 kΩ
Bit 8   2 kΩ
Bit 9   1 kΩ

In order to ensure high accuracy, all of these resistors would need to be close-tolerance types (typically ±1%, or better). A more practical arrangement uses an operational amplifier in which the input voltage to the operational amplifier is determined by means of an R-2R ladder, as shown in Figure 3.6(c). Note that only two resistance values are required and that they can be any convenient value provided that one value is double the other (it is relatively easy to manufacture matched resistances of close tolerance and high-stability on an integrated circuit chip).

3.2.1 Resolution and accuracy

The **accuracy** of a DAC depends not only on the values of the resistance used also on the reference voltage used to define the voltage levels. Special **band-gap references** (similar to precision zener diodes) are normally used to provide reference voltages that are closely maintained over a wide range of temperature and supply voltages. Typical accuracies of between 1% and 2% can be achieved using most modern low-cost DAC devices.

The **resolution** of a DAC is an indication of the number of increments in output voltage that it can produce and it is directly related to the number of binary digits used in the conversion. The two simple four-bit DACs that we met earlier can each provide sixteen different output voltages but in practice we would probably require many more (and correspondingly smaller) increments in output voltage. This can be achieved by adding further binary inputs.

For example, a DAC with eight inputs (i.e. an 8-bit DAC) would be capable of producing 256 (i.e. 2^8) different output voltage levels. A 10-bit device, on the other hand, will produce 1,024 (i.e. 2^{10}) different voltage levels. The resolution of a DAC is generally stated in terms of the number of binary digits (i.e. bits) used in the conversion.

Key Point

The resolution of a DAC depends on the number of bits used in the conversion process—the more bits the greater the resolution. Typical DAC have resolutions of 8, 10 or 12 bits.

3.2.2 Filters

As we have seen, the output of a DAC consists of a series of quantized voltage levels. The presence of these levels on the output signal can be undesirable for some applications and hence they are removed in order to 'smooth' the output voltage. This can be easily accomplished by passing the output signal through a low-pass filter, as shown in Figure 3.7. The filter is designed so that the residual sampling frequency components (i.e. those that cause the 'steps' in the analogue signal) are well beyond the cut-off frequency of the filter and are therefore subject to an appreciable amount of attenuation.

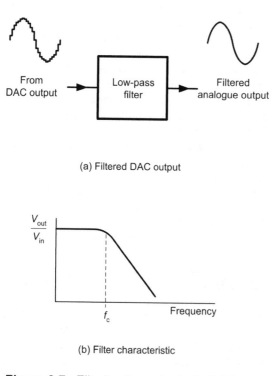

(a) Filtered DAC output

(b) Filter characteristic

Figure 3.7   Filtering the output of a DAC

3.3 Analogue to digital conversion

The basic analogue to digital converter (ADC) has a single analogue input and a number of digital outputs (often 8, 10, 12, or 16 lines), as shown in Figure 3.8.

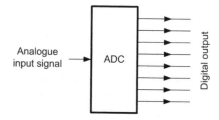

Figure 3.8   Basic ADC arrangement

Various forms of analogue to digital converter are available for use in different applications including multi-channel ADC with up to 16 analogue inputs. The simplest form of ADC is the **flash converter** shown in Figure 3.9(a). In this type of ADC the incoming analogue voltage is compared with a series of fixed reference voltages using a number of operational amplifiers (IC1 to IC7 in Figure 3.9). When the analogue input voltage exceeds the reference voltage present at the inverting input of a particular operational amplifier stage the output of that stage will go to logic 1. So, assuming that the analogue input voltage is 2V, the outputs of IC1 and IC2 will go to logic 1 whilst the remaining outputs will be at logic 0.

The **priority encoder** is a logic device that produces a binary output code that indicates the value of the most significant logic 1 received on one of its inputs. In this case, the output of IC2 will be the most significant logic 1 and hence the binary output code generated will be 010 as shown in Figure 3.9(b). Flash ADC are extremely fast in operation (hence the name) but they become rather impractical as the resolution increases. For example, an 8-bit flash ADC would require 256 operational amplifier comparators and a 10-bit device would need a staggering 1,024 comparator stages! Typical conversion times for a flash ADC lie in the range 50 ns to 1 μs so this type of ADC is ideal for 'fast' or rapidly changing analogue signals. Due to their complexity, flash ADC are relatively expensive.

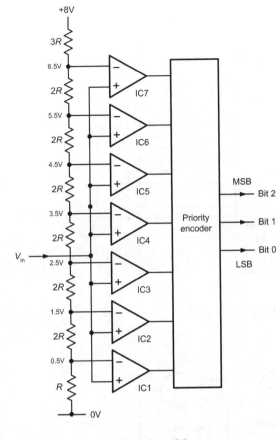

Input	Output		
	Bit 2	Bit 1	Bit 0
$V_{in} < 0.5V$	0	0	0
$0.5V < V_{in} < 1.5V$	0	0	1
$1.5V < V_{in} < 2.5V$	0	1	0
$2.5V < V_{in} < 3.5V$	0	1	1
$3.5V < V_{in} < 4.5V$	1	0	0
$4.5V < V_{in} < 5.5V$	1	0	1
$5.5V < V_{in} < 6.5V$	1	1	0
$V_{in} > 6.5V$	1	1	1

(b) Truth table

(a) Simple flash ADC

Figure 3.9   A simple flash ADC

A **successive approximation ADC** is shown in Figure 3.10. This shows an 8-bit converter that uses a DAC (usually based on an R-$2R$ ladder) together with a single operational amplifier comparator (IC1) and a **successive approximation register (SAR)**. The 8-bit output from the SAR is applied to the DAC and to an 8-bit output latch. A separate **end of conversion (EOC)** signal (not shown in Fig. 3.10) is generated to indicate that the conversion process is complete and the data is ready for use.

When a **start conversion (SC)** signal is received, successive bits within the SAR are set and reset according to the output from the comparator. At the point at which the output from the comparator reaches zero, the analogue input voltage will be the same as the analogue output

from the DAC and, at this point, the conversion is complete. The end of conversion signal is then generated and the 8-bit code from the SAR is read as a digital output code.

Successive approximation ADC are significantly slower than flash types and typical conversion times (i.e. the time between the SC and EOC signals) are in the range 10 μs to 100 μs. Despite this, conversion times are fast enough for most non-critical applications and this type of ADC is relatively simple and available at low-cost.

A **ramp-type ADC** is shown in Fig. 3.11. This type of ADC uses a ramp generator and a single operational amplifier comparator, IC1. The output of the comparator (either a 1 or a 0 depending on whether the input voltage is greater or less than

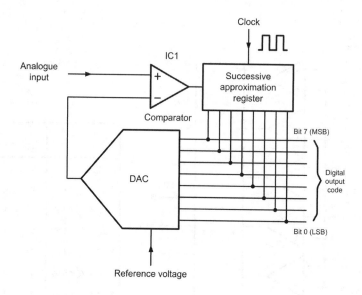

Figure 3.10 A successive approximation ADC

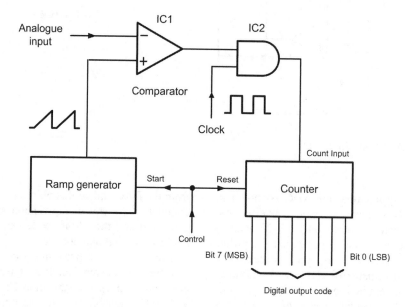

Figure 3.11 A ramp-type ADC

the instantaneous value of the ramp voltage). The output of the comparator is used to control a logic gate (IC2) which passes a clock signal (a square wave of accurate frequency) to the input of a pulse counter whenever the input voltage is greater than the output from the ramp generator. The pulses are counted until the voltage from the ramp generator exceeds that of the input signal, at which point the output of the comparator goes low and no further pulses are passed into the counter. The number of clock pulses counted will depend on the input voltage and the final binary count can thus provide a digital representation of the analogue input. Typical waveforms for the ramp-type waveform are shown in Fig. 3.12.

Finally, the **dual-slope ADC** is a refinement of the ramp-type ADC which involves a similar comparator arrangement but uses an internal voltage reference and an accurate fixed slope negative ramp which starts when the positive going ramp reaches the analogue input voltage. The important thing to note about this type of

ADC is that, whilst the slope of the positive ramp depends on the input voltage, the negative ramp falls at a *fixed* rate. Hence this type of ADC can provide a very high degree of accuracy and can also be made so that it rejects noise and random variations present on the input signal. The main disadvantage, however, is that the process of ramping up and down requires some considerable time and hence this type of ADC is only suitable for 'slow' signals (i.e. those that are not rapidly changing). Typical conversion times lie in the range 500 μs to 20 ms.

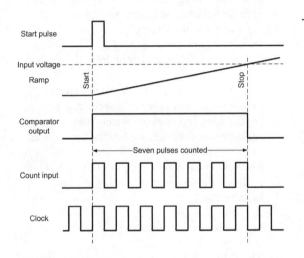

Figure 3.12 Waveforms for a single-ramp ADC

Test your understanding 3.1

The binary codes produced by a four-bit bipolar analogue to digital converter (see Figs. 3.3 and 3.4) and sampled at intervals of 1 ms, have the following values:

Time (ms)	Binary code
0	0101
1	0100
2	0011
3	0010
4	0001
5	0000
6	1111
7	1110

Sketch the waveform of the analogue voltage and name its shape.

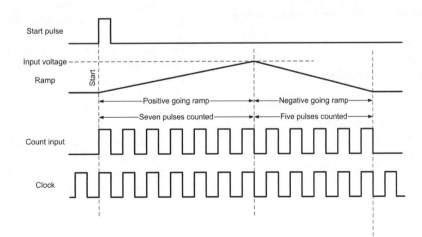

Figure 3.13 Waveforms for a dual-ramp ADC

Test your understanding 3.2

What type of ADC would be most suitable for each of the following applications? Give reasons for your answers.

1. Sensing and recording the strain in a beam to an accuracy of better than 1% and with a resolution of 1 part in 10^3.

2. Converting a high-quality audio signal into a digital data stream for recording on a CD-ROM.

3. Measuring DC voltages (that may be accompanied by supply-borne hum and noise) in a digital voltmeter.

3.4 Multiple choice questions

1. A DAC can produce 256 different output voltages. This DAC has a resolution of:
   (a) 8 bits
   (b) 128 bits
   (c) 256 bits.

2. In a bipolar ADC a logic 1 in the MSB position indicates:
   (a) zero input voltage
   (b) negative input voltage
   (c) positive input voltage.

3. Which one of the following types of ADC is the fastest?
   (a) ramp type
   (b) flash type
   (c) successive approximation type.

4. Which one of the following ADC types uses a large number of comparators?
   (a) ramp type
   (b) flash type
   (c) successive approximation type.

5. The advantage of an R-$2R$ ladder DAC is that:
   (a) it uses less resistors overall
   (b) it uses a smaller number of analogue switches
   (c) it avoids having a large number of different value resistors.

6. A 10-bit DAC is capable of producing:
   (a) 10 different output levels
   (b) 100 different output levels
   (c) 1,024 different output levels.

7. The process of sampling approximating an analogue signal to a series of discrete levels is referred to as:
   (a) interfacing
   (b) quantizing
   (c) data conversion.

8. In a binary weighted DAC the voltage gain for each digital input is determined by:
   (a) a variable slope ramp
   (b) a variable reference voltage
   (c) resistors having different values.

9. The resolution of a DAC is stated in terms of:
   (a) the accuracy of the voltage reference
   (b) the open-loop gain of the comparator
   (c) the number of bits used in the conversion.

10. The conversion time of a flash ADC is typically in the range:
    (a) 50 ns to 1 μs
    (b) 50 μs to 1 ms
    (c) 50 ms to 1 s.

11. The DAC used in a successive approximation ADC is usually:
    (a) ramp type
    (b) binary weighted type
    (c) successive approximation type.

12. In a successive approximation ADC, the time interval between the SC and EOC signals is:
    (a) the clock time
    (b) the cycle time
    (c) the conversion time.

13. An advantage of a dual-ramp ADC is:
    (a) a fast conversion time
    (b) an inherent ability to reject noise
    (c) the ability to operate without the need for a clock.

<table>
<tr><td>

Chapter

4

</td><td>

Data buses

</td></tr>
</table>

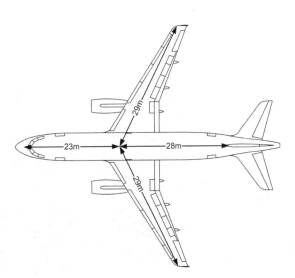

Aircraft data bus systems allow a wide variety of avionics equipment to communicate with one another and exchange data. In this section we shall take a brief look at the principles of aircraft data bus systems before introducing some of the systems that are commonly used in modern aircraft.

4.1 Introducing bus systems

The word 'bus' is a contraction of the Greek word 'omnibus' and the word simply means 'to all'. Thus, in the context of computers and digital systems, 'bus' refers to a system that permits interconnection and data exchange between the devices in a complex system. Note, however that 'interconnection' involves more than just physical wiring, amongst other things it defines the voltage levels and rules (or **protocols**) that govern the transfer of data.

   With such a large number of avionic systems, a modern aircraft requires a considerable amount of cabling. Furthermore, some of the cabling runs in a large aircraft can be quite lengthy, as shown in Figure 4.1. Aircraft cabling amounts to a significant proportion of the unladen weight of an aircraft and so minimising the amount of cabling and wiring present is an important consideration in the design of modern aircraft, both civil and military.

4.1.1 Bus terminology

Bus systems can be either **bidirectional** (one way) or **unidirectional** (two way), as shown in Figure 4.2. They can also be **serial** (one bit of data transmitted at a time) or **parallel** (where often 8, 16 or 32 bits of data appear as a group on a number of data lines at the same time). Because of the constraints imposed by conductor length and weight, all practical aircraft bus systems are based on serial (rather than parallel) data transfer.

Figure 4.1 Typical dimensions of a modern passenger aircraft

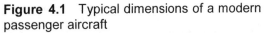

(a) Unidirectional serial data

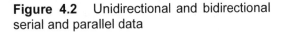

(b) Bidirectional serial data

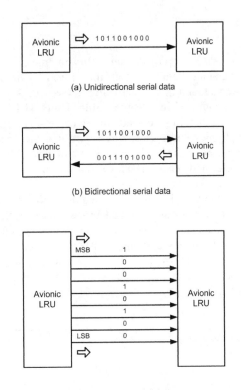

(b) Unidirectional parallel data

Figure 4.2 Unidirectional and bidirectional serial and parallel data

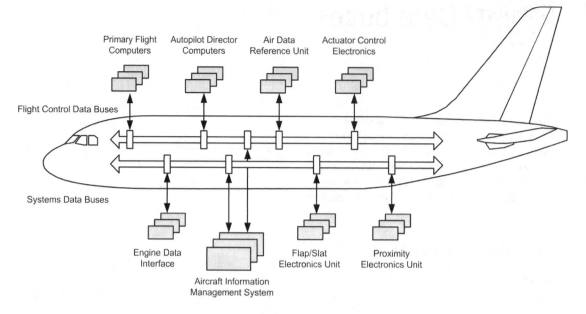

Primary Flight   Autopilot Director   Air Data   Actuator Control
Computers   Computers   Reference Unit   Electronics

Flight Control Data Buses

Systems Data Buses

Engine Data   Flap/Slat   Proximity
Interface   Electronics Unit   Electronics Unit

Aircraft Information
Management System

Figure 4.3   Multiple bus systems implemented on a modern passenger aircraft

Bus systems provide an efficient means of exchanging data between the diverse avionic systems found in a modern aircraft (see Fig. 4.3).

Individual **Line Replaceable Units (LRU)**, such as the Engine Data Interface or Flap/Slat Electronics Units shown in Figure 4.3, are each connected to the bus by means of a dedicated **bus coupler** and **serial interface module** (not shown in Fig. 4.3).

Within the LRU, the dedicated digital logic and microprocessor systems that process data locally each make use of their own **local bus** system. These local bus systems invariably use parallel data transfer which is ideal for moving large amounts of data very quickly but only over short distances.

Key Point

Modern aircraft use multiple redundant bus systems for exchanging data between the various avionic systems and sub-systems. These bus systems use serial data transfer because it minimises the size and weight of aircraft cabling.

4.1.2 Bus protocols

The notion of a protocol needs a little explaining so imagine for a moment that you are faced with the problem of organizing a discussion between a large number of people sitting around a table who are blindfolded and therefore cannot see one another. In order to ensure that they didn't all speak at once, you would need to establish some ground rules, including how the delegates would go about indicating that they had something to say and also establishing some priorities as to who should be allowed to speak in the event that several delegates indicate that they wish to speak at the same time. These (and other) considerations would form an agreed protocol between the delegates for conducting the discussion. The debate should proceed without too many problems provided that everybody in the room understands and is willing to accept the protocol that you have established. In computers and digital systems **communications protocols** are established to enable the efficient exchange of data between multiple devices connected to the same bus. A number of different standards are commonly used.

4.1.3 Bus architecture

Bus architecture is a general term that refers to the overall structure of a computer or other digital system that relies on a bus for its operation. The architecture is often described in the form of a block schematic diagram showing how the various system elements are interconnected and also how the data flow is organised between the elements. The architecture of a system based on the use of a unidirectional serial bus system is shown in Figure 4.4(a) whilst a comparable bi-directional bus system is shown in Figure 4.4(b). Note how the bidirectional system simplifies the interconnection of the LRUs and allows all of the devices to transmit and receive on the same bus.

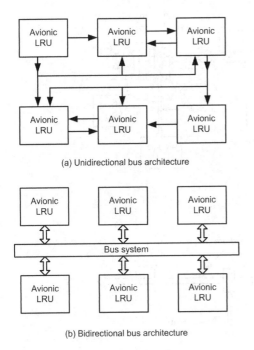

(a) Unidirectional bus architecture

(b) Bidirectional bus architecture

Figure 4.4 Bus architecture

Key Point

Communication protocols enable the efficient exchange of data between a number of devices connected to the same bus. Protocols consist of a set of rules and specifications governing, amongst other things, data format and physical connections.

4.1.4 Serial bus principles

A simple system for serial data transfer between two Line Replaceable Units (LRU) each of which comprises an avionic system in its own right is shown in Figure 4.5. Within the LRU data is transferred using an internal parallel data bus (either 8, 16, 32 or 64 bits wide). The link between the two LRUs is made using a simple serial cable (often with only two, four or six conductors). The required parallel-to-serial and serial-to-parallel data conversion is carried out by a bus interface (often this is a single card or module within the LRU). The data to be transferred can be **synchronous** (using clock signals generated locally within each LRU) or it may be **asynchronous** (i.e. self-clocking).

The system shown in Fig. 4.5 has the obvious limitation that data can only be exchanged between two devices. In practice we need to share the data between many LRU/avionic units. This can be achieved by the bus system illustrated in Fig. 4.6. In this system, data is transferred using a **shielded twisted pair (STP) bus cable** with a number of **coupler panels** that are located at appropriate points in the aircraft (e.g. the flight deck, avionics bay, etc). Each coupler panel allows a number of avionic units to be connected to the bus using a **stub cable**. In order to optimise the speed of data transfer and minimise problems associated with reflection and mismatch, the bus cable must be terminated at each end using a matched **bus terminator**.

Bus couplers are produced as either **voltage mode** or **current mode** units depending upon whether they use voltage or current sensing devices. Within each LRU/avionics unit, an interface is provided that performs the required serial-to-parallel or parallel-to-serial data conversion, as shown in Figure 4.7.

As well as providing an electrical interface (with appropriate voltage and current level shifting) the interface unit also converts the data formats (e.g. from serial analogue doublets present in the stub cable to Manchester-encoded serial data required by the Terminal Controller) as shown in Figure 4.8.

In order to transmit data using the serial data bus information must be presented in a standard format. A typical format for serial data would use

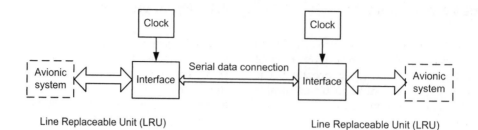

Figure 4.5 A simple system for serial data transfer between two avionic systems

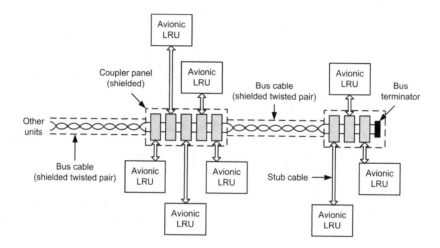

Figure 4.6 A practical aircraft data bus

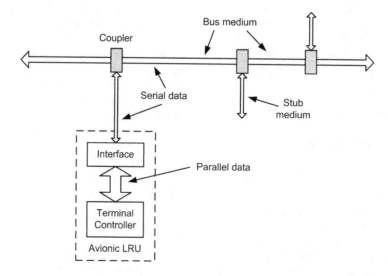

Figure 4.7 A basic bus interface

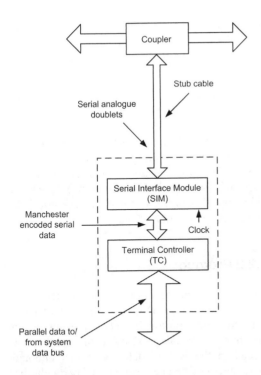

Figure 4.8 Data formats in a practical bus interface

a word length of 32 bits. This word comprises of several discrete fields including:

- Up to 20 bits for data (which may be further divided)
- An 8-bit label field which is used to identify the data type and any parameters that may be associated with it
- A source/destination identifier (SDI).
- A number of status bits used to provide information about the mode, hardware condition, or validity of data
- An added parity bit which provides a means of validating the data (i.e. determining whether or not it is free from error).

Key Point

A means of converting serial data to parallel data (and vice versa) is required whenever an LRU is to be interfaced to an aircraft bus system.

4.2 ARINC 429

The ARINC 429 data bus has proved to be one of the most popular bus standards used in commercial aircraft. The ARINC 429 specification defines the electrical and data characteristics and protocols that are used.

ARINC 429 employs a unidirectional data bus standard known as Mark 33 Digital Information Transfer System (DITS). Messages are transmitted in packets of 32-bits at a bit rate of either 12.5 or 100 kilobits per second (referred to as low and high bit rate respectively). Because the bus is unidirectional, separate ports, couplers and cables will be required when an LRU wishes to be able to both transmit and receive data. Note that a large number of bus connections may be required on an aircraft that uses sophisticated avionic systems.

ARINC 429 has been installed on a wide variety of commercial transport aircraft including; Airbus A310/A320 and A330/A340; Boeing 737, 747, 757, and 767; and McDonnell Douglas MD-11. More modern aircraft (e.g. Boeing 777 and Airbus A380) use significantly enhanced bus specifications (see page 40) in order to reduce the weight and size of cabling and to facilitate higher data rates than are possible with ARINC 429. Despite these moves to faster, bidirectional bus standards, the ARINC 429 standard has proved to be highly reliable and so is likely to remain in service for many years to come.

4.2.1 Electrical characteristics

ARINC 429 is a two wire differential bus which can connect a single transmitter or source to one or more receivers or sinks. Two speeds are

Key Point

Aeronautical Radio Inc. (ARINC) is an organization composed of major airlines and aircraft manufacturers which seeks to promote standardization within aircraft equipment. More information on ARINC and aircraft standards can be obtained from www.arinc.com

available, 12.5 k bps (bits per second) and 100 kbps. The data bus uses two signal wires to transmit 32-bit words. Transmission of sequential words is separated by at least four bit times of NULL (zero voltage). This eliminates the need for a separate clock signal and it makes the system **self-clocking**.

The ARINC 429 electrical characteristics are summarised below:

Voltage levels:	+5V, 0V, −5V (each conductor with respect to ground) +10V, 0V, −10V (conductor A with respect to conductor B)
Data encoding:	Bi-Polar Return to Zero
Word size:	32 bits
Bit rate (high):	100K bits per second
Bit rate (low):	12.5K bits per second
Slew rate (high):	1.5ms (±0.5 ms)
Slew rate (low):	10ms (±5 ms)

The nominal transmission voltage is 10V ±1V between wires (differential), with either a positive or negative polarity. Therefore, each signal leg ranges between +5V and −5V. If one conductor is at +5V, the other is conductor is at −5V and vice versa. One wire is called the 'A' (or '+' or 'HI') conductor and the other is called the 'B' (or '−' or 'LO') wire. The modulation employed is **bipolar return to zero** (BPRZ) modulation (see Fig. 4.9). The composite signal state may be at one of the following three levels (measured between the conductors):

- **HI** which should be within the range +7.25V to 11V (A to B)
- **NULL** which should be within the range +0.5V to −0.5V (A to B)
- **LO** which should be within the range −7.25V to −11V (A to B).

The received voltage on a serial bus depends on line length and the number of receivers connected to the bus. With ARINC 429, no more than 20 receivers should be connected to a single bus. Since each bus is unidirectional, a system needs to have its own transmit bus if it is required to respond to or to send messages. Hence, to achieve bidirectional data transfer it is necessary to have two separate bus connections.

4.2.2 Protocol

Since there can be only one transmitter on a twisted wire pair, ARINC 429 uses a very simple, point-to-point protocol. The transmitter is continuously sending 32-bit data words or is placed in the NULL state. Note that although there may only be one receiver on a particular bus cable the ARINC specification supports up to 20.

4.2.3 Bit timing and slew rate

The slew rate refers to the rise and fall time of the ARINC waveform. Specifically, it refers to the amount of time it takes the ARINC signal to rise from the 10% to the 90% voltage amplitude points on the leading and trailing edges of a

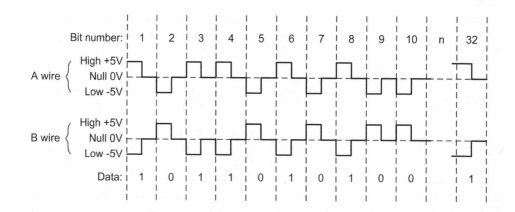

Figure 4.9   Signals present on the twisted pair conductors in the ARINC 429 aircraft data bus

pulse. The data shown in Table 4.1 applies to the high and low-speed ARINC 429 systems.

Table 4.1 ARINC 429 parameters

Parameter	High-speed	Low-speed
Bit rate	100K bps	12.5K to 14.5K bps
Bit time (Y)	10 μs ± 5%	1/bit rate μs ± 5%
High time (X)	5 μs ± 5%	Y/2 μs ± 5% (see Fig. 4.10)
Rise time	1.5 μs ± 0.5μs	10 μs ± 5 μs
Fall time	1.5 μs ± 0.5μs	10 μs ± 5 μs

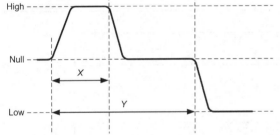

Figure 4.10 Timing diagram for logic level transitions in the ARINC429 data bus

Test your understanding 4.1

1. Explain the difference between serial and parallel methods of data transfer.

2. Explain why serial data transfer is used for aircraft bus systems.

3. Explain the function of each of the following bus system components:
   (a) Bus cable
   (b) Stub cable
   (c) Bus terminator
   (d) Coupler panel.

4. State the voltage levels present on the twisted pair conductors in an ARINC 429 data bus.

4.2.4 ARINC 429 data word format

In most cases, an ARINC message consists of a single 32-bit data word (see Fig. 4.11). The 8-bit label field defines the type of data that is contained in the rest of the word. ARINC data words are always 32 bits and typically include five primary fields, namely Parity, SSM, Data, SDI, and Label. ARINC convention numbers the bits from 1 (LSB) to 32 (MSB). A number of different data formats is possible.

32	31 30	29	11	10 9	8	1
P	SSM	MSB DATA LSB		SDI	LABEL	

Figure 4.11 Basic ARINC 429 data word format (note the total length is 32 bits)

Bits are transmitted starting with bit 1 of the label and the final bit transmitted is the parity bit. The standard specifies the use of **odd parity** (the parity bit is set to 1 or reset to 0 in order to ensure that there is an odd number of 1s in each transmitted word). It is worth noting that the label is transmitted with the most significant bit (MSB) first while the data is transmitted least significant bit (LSB) first.

The **Label field** is an octal value that indicates the type of data (e.g. airspeed, altitude, etc) that is being transmitted.

The **SDI field** is used when a transmitter is connected to multiple receivers but not all data is intended for used by all the receivers. In this case each receiver will be assigned an SDI value and will look only at labels which match its SDI value. While the specification calls for SDI 00 to be universally accepted this may not actually be the case.

The **Data field** contains the actual data to be sent. The principal data formats defined in the specification are Binary Coded Decimal (BCD) which uses groups of four bits to contain a single decimal digit and BNR which is binary coding. For both of these data types, the specification defines the units, the resolution, the range, the number of bits used and how frequently the label should be sent.

The **SSM field** is used for information which assists the interpretation of the numeric value in the data field. Examples of SSM values might be North, East, South, West, Plus, Minus, Above or Below.

The **P field** is the parity bit. ARINC 429 uses odd parity. The parity bit is the last bit transmitted within the data word.

Some examples of data sent over an ARINC 429 bus are shown in Figure 4.12 and 4.13. In Figure 4.12 a **BCD** word (see page 17) is being transmitted whilst in Figure 4.13 the data is encoded in binary format. The ARINC 429 binary specification calls for the use of two's complement notation to indicate negative numbers (see page 18) and this binary format is known as **BNR**.

In Figure 4.13 the label (103) corresponds to Selected Airspeed and the indicated value is 268 knots (256 + 8 + 4). The zero in bit-29 position indicates a positive value and the data (presented in natural binary format) uses bit-28 (for the MSB) to bit-20 (for the LSB). The remaining bits are padded with zeros. In Figure 4.12 the BNR

data conveys a value (this time expressed in BCD format) of 25786.

Tables 4.2 and 4.3 provide some examples of labels and equipment ID (SDI).

Test your understanding 4.2

1. Explain the difference between BNR and BCD encoding.

2. Using BNR encoding, which bit in an ARINC 429 data word indicates whether the data is positive or negative?

3. What is the largest positive value that can be transmitted using a single ARINC 429 data word using:
   (a) BNR encoding
   (b) BCD encoding.

4. In relation to an ARINC 429 data word, explain the function of:
   (a) the SDI field
   (b) the SSM field.

32	31	30	29	28	27	26	25	24	23	22	21	20	19	18	17	16	15	14	13	12	11	10	9	8	1
P	SSM	CHAR 1			CHAR 2			CHAR 3			CHAR 4			CHAR 5			SDI			LABEL					

(a) BCD word format

32	31	30	29	28	27	26	25	24	23	22	21	20	19	18	17	16	15	14	13	12	11	10	9	8	1
P	SSM	0	1	0	0	1	0	1	0	1	1	1	1	0	0	0	0	0	1	1	0	SDI		LABEL	
	0	0	2			5			7			8			6										

(b) BCD word example

Figure 4.12   ARINC 429 BCD data word format

32	31	30	29	28	27	26	25	24	23	22	21	20	19	18	17	16	15	14	13	12	11	10	9	8	1
P	SSM	DATA																		PAD		SDI		LABEL	

(a) BNR encoding

32	31	30	29	28	27	26	25	24	23	22	21	20	19	18	17	16	15	14	13	12	11	10	9	8	1
P	SSM	DATA																		PAD		SDI		LABEL	
0	1	1	0	1	0	0	0	0	1	1	0	0	0	0	0	0	0	0	0	0	0	0	0	103	

(b) BNR encoding example

Figure 4.13   Basic ARINC 429 BNR word format (note the total length is 32 bits)

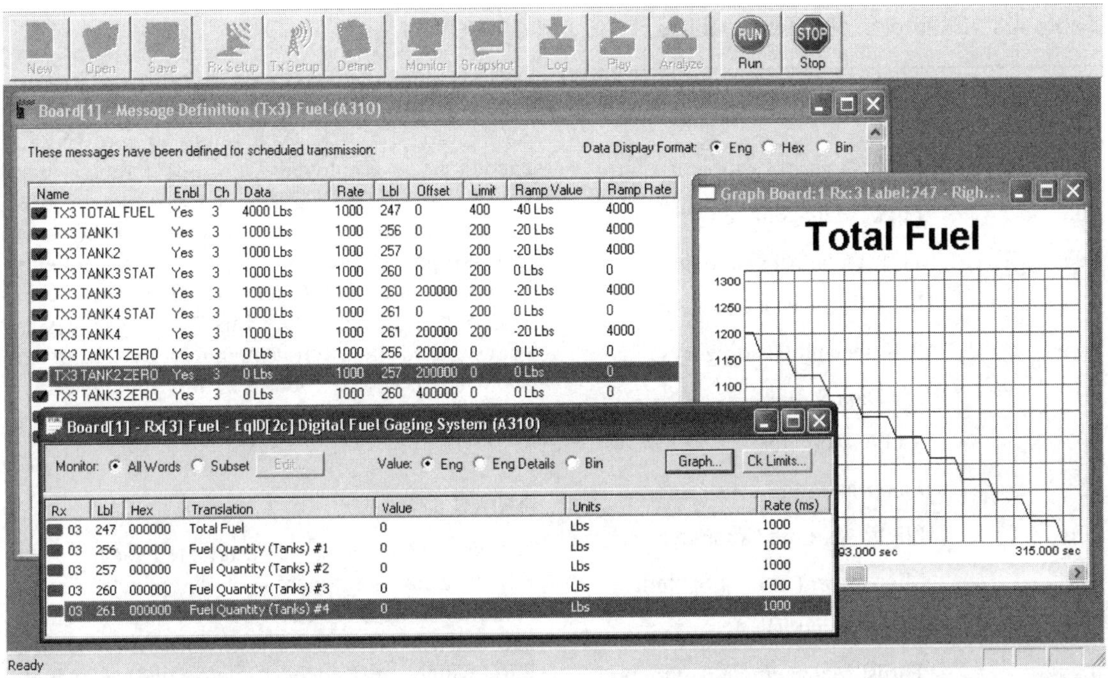

Figure 4.14 Using a bus analyzer (Bus Tools from Condor Engineering) to examine the data present on an ARINC 429 data bus and display the rate at which fuel is being used

Table 4.2 Examples of ARINC 429 label codes

Label (octal)	Transmitted code (binary)	Equip. ID (hex)	Parameter transmitted	Data format
140	01 100 000	001	Flight Director – Roll	BNR
		025	Flight Director – Roll	BNR
		029	Pre-cooler Output Temperature	BNR
		05A	Actual Fuel Quantity Display	BCD
141	01 100 001	001	Flight Director – Pitch	BNR
		025	Flight Director – Pitch	BNR
		029	Pre-cooler Input Temperature	BNR
		05A	Pre-selected Fuel Quantity Display	BCD
142	01 100 010	002	Flight Director – Fast/Slow	BNR
		003	Flight Director – Fast/Slow	BNR
		025	Flight Director – Fast/Slow	BNR
		05A	Left Wing Fuel Quantity Display	BCD

Table 4.3 Examples of equipment IDs

Equipment ID (hex)	Equipment type
001	Flight Control Computer
002	Flight Management Computer
003	Thrust Control Computer
004	Inertial Reference System
005	Attitude and Heading System
006	Air Data System
007	Radio Altimeter
025	Electronic Flight Instruments
026	Flight Warning Computer
027	Microwave Landing System
029	ADDCS and EICAS
02A	Thrust Management Computer

4.3 Other bus standards

The following is a brief summary of some of the other aircraft data bus systems that have appeared over the last forty years. It is important to note that these often describe enhancements to existing standards. However, in all cases, the main aim is that of ensuring that equipment manufacturers and operators are working to a common specification which ensures that hardware and software is both interoperable and upgradeable.

ARINC 419

The ARINC 419 standard describes several digital transmission standards that predate ARINC 429. Some of these used 32-bit words similar to ARINC 429. Some standards were based on the use of a six wire system whilst others used a shielded two wire twisted pair (like ARINC 429) or a coaxial cable. Line voltage levels were either two state (HI/LO) or three state (HI/NULL/LO) with voltages ranging from 10 V to 18.5 V for the high state and from less than 1 V to less than 5 V for the NULL state.

ARINC 561

ARINC 561 was based on a six wire system involving three pairs that were used for DATA, SYNC, and CLOCK. Non return to zero (NRZ) encoding was employed with logic 1 represented by 12 V. Like ARINC 429, the word length was 32 bits with bits 32 and 31 comprising the SSM and no parity available. The remaining fields include an 8-bit label and 6 BCD fields, five of four bits and one of two bits. This system was widely used in aircraft manufactured prior to about 1970. ARINC 568 uses the same electrical interface specification as used in ARINC 561.

ARINC 573

ARINC 573 is the standard adopted for use with Flight Data Recorders (FDR) which use a continuous data stream of Harvard Bi-Phase encoded 12-bit words. These words are encoded into which are encoded into **frames** which contain a snapshot of the data from each of the avionics subsystems on the aircraft. Each frame comprises four sub-frames. A unique synchronising word appears at the start of each sub-frame. **ARINC 717** supersedes ARINC 573 and caters for a number of different bit rates and frame sizes.

ARINC 575

Similar to ARINC 429, this standard is a low speed bus system that is based on a single twisted pair of wires. Due to the low data rate supported, this bus standard is now considered obsolete. Electrically, ARINC 575 is generally compatible with low speed ARINC 429. However, some variants of ARINC 575 use a bit rate that is significantly slower than ARINC 429 and they may not be compatible in terms of the electrical specification and data formats.

ARINC 615

ARINC 615 is a **software protocol** that can be layered on top of ARINC 429 compatible systems. ARINC 615 supports high speed data transfer to and from on-board digital systems permitting, for example, reading and writing of 3½ inch disks.

ARINC 629

ARINC 629 was introduced in the mid 1990s and it supports a data rate of 2 Mbps (20 times faster than ARINC 429). The bus supports 120 connected devices and is currently used on the Boeing 777, Airbus A330 and A340 aircraft. A notable enhancement of the earlier ARINC 429 standard is that ARINC 629 is a bi-directional bus system (in other words, connected devices can transmit, receive or do both). Another advantage of ARINC 629 is that it achieves bidirectional bus communication without the need for a bus controller (which could be a potential source of single-point failure). The physical bus medium is shielded twisted pair (STP).

ARINC 708

ARINC 708 is used to transfer data from the airborne weather radar receiver to the aircraft's radar display. The bus is unidirectional and uses Manchester encoded data at a data rate of 1 Mbps. Data words are 1600 bits long and they are composed of one 64-bit status word and 512, 3-bit data words.

MIL-STD-1553B/1773B

Military standard 1533B is a bidirectional centrally controlled data bus designed for use in military aircraft. The standard uses a Bus Controller (BC) which can support up to 31 devices which are referred to as Remote Terminals (RT). The standard supports a bit rate of 1 Mbps. MIL-STD-1773B is a fibre optic implementation of MIL-STD-1553B that provides significantly greater immunity to exposure to high-intensity radiated electromagnetic fields (HIRF).

CSDB and ASCB

The CSDB and ASCB standards are proprietary protocols from Collins and Honeywell respectively. These systems are often used in small business and private general aviation (GA) aircraft. CSDB is a unidirectional bus that permits the connection of up to 10 receivers and one transmitter. The standard supports data rates of 12.5 kbps and 50 kbps. ASCB is a centrally controlled bidirectional bus. A basic configuration comprises a single Bus Controller and two isolated buses, each of which can support up to 48 devices.

FDDI

The Fibre Distributed Data Interface (FDDI) was originally developed by Boeing for use on the Boeing 777 aircraft. FDDI is a local area network (LAN) based on a dual token ring topology. Data in each ring flows in opposite directions. The data rate is 100 Mbps and data is encoded into frames. CDDI and SDDI are similar network bus standards based on copper (Copper Distributed Data Interface) and shielded twisted pair (SDDI) as the physical media. The data format is NRZI (a data format similar to NRZ but where a change in the line voltage level indicates a logic 1 and no change indicates a logic 0). For reasons of cost and in order to reduce the number and complexity of network standards used in its aircraft, Boeing now plans to replace the system on the 777 with a less-expensive 10 Mbps copper Ethernet.

Test your understanding 4.3

Which of the listed bus standards would be most suitable for each of the following applications? Give reasons for your answers.

1. A low cost bus system for the simple avionics fitted to a small business aircraft.

2. Connecting a weather radar receiver to a radar display.

3. A bus system for linking the various avionics systems of a modern passenger aircraft to its flight data recorder (FDR)..

4. A bus system for the avionics of a military aircraft fitted with multiple radars and electronic counter measures (ECM).

4.4 Multiple choice questions

1. A bus that supports the transfer of data in both
   directions is referred to as:
   (a) universal
   (b) bidirectional
   (c) asynchronous.

2. The main advantage of using a serial bus in an
   aircraft is:
   (a) there is no need for data conversion
   (b) it supports the highest possible data rates
   (c) reduction in the size and weight of cabling.

3. Which one of the following is used to
   minimize reflections present in a bus cable?
   (a) coupler panels
   (b) bus terminators
   (c) shielded twisted pair cables.

4. The data format in an ARINC 429 stub cable
   consists of:
   (a) serial analogue doublets
   (b) parallel data from the local bus
   (c) Manchester encoded serial data.

5. In order to represent negative data values,
   BNR data uses:
   (a) two's complement binary
   (b) BCD data and a binary sign bit.
   (c) an extra parity bit to indicate the sign

6. The physical bus media specified in ARINC
   629 is:
   (a) fiber optic
   (b) coaxial cable
   (c) shielded twisted pair.

7. The validity of an ARINC 429 data word is
   checked by using:
   (a) a parity bit
   (b) a checksum
   (c) multiple redundancy.

8. The label field in an ARINC 429 data word
   consists of:
   (a) five bits
   (b) three bits
   (c) eight bits.

9. The voltages present on an ARINC 429 data
   bus cable are:
   (a) ±5 V
   (b) ±15 V
   (c) +5 V and +10V

10. The maximum bit rate supported by ARINC
    429 is:
    (a) 12.5 Kbps
    (b) 100 Kbps
    (c) 1 Mbps.

11. An ARINC 429 NULL state is represented by:
    (a) a voltage of −5 V
    (b) a voltage of 0 V
    (c) a voltage of +5 V.

12. The maximum data rate supported MIL-STD-
    1553 is:
    (a) 12.5 Kbps
    (b) 100 Kbps
    (c) 1 Mbps.

13. A bus that is self-clocking is referred to as:
    (a) universal
    (b) bidirectional
    (c) asynchronous.

14. The physical bus media specified in MIL-
    STD-1773B is:
    (a) fiber optic
    (b) coaxial cable
    (c) shielded twisted pair.

15. The length of an ARINC 429 word is:
    (a) 16 bits
    (b) 20 bits
    (c) 32 bits.

16. ARINC 573 is designed for use with:
    (a) INS
    (b) FDR
    (c) Weather radar.

17. The maximum data rate supported by the
    FDDI bus is:
    (a) 1 Mbps
    (b) 10 Mbps
    (c) 100 Mbps.

Logic circuits

An ability to make decisions based on a variety of different factors is crucial to the safe operation of a modern aircraft. It is therefore not surprising that logic systems are widely used in aircraft today. In this chapter we introduce the basic building blocks of digital logic circuits (AND, OR, NAND, NOR, etc.) together with the symbols and truth tables that describe the operation of the most common logic gates. We then show how these gates can be used in simple combinational logic circuits before moving on to introduce bistable devices, counters and shift registers. The chapter concludes with a brief introduction to the two principal technologies used in modern digital logic circuits, TTL and CMOS.

5.1 Introducing logic

We all make decisions based on logic in our everyday lives. The sort of decisions that we make are invariably conditional (i.e. they depend on a particular set of circumstances) and they often take the form:

If {condition} then {action}

For example:

If cold then put on the heating

or

If thirsty then go to the bar

We also make compound decisions that are based on more than one set of circumstances and can take the form:

If {condition 1} or {condition 2} then {action}

or

If {condition 1} and {condition 2} then {action}

For example:

If hungry or thirsty then go to the cafe

or

If cold and dark then go to bed

Another possibility is that we might want to make a decision based on the absence of a condition rather than its presence. For example:

If {condition 1} or {condition 2} and not {condition 3} then {action}

An example of this might be:

If hungry or thirsty and not raining then walk to the pub

The important words in the above conditional statements are 'and', 'or' and 'not'. These words describe logical conditions and we will next look at ways in which electronic circuits can emulate these **logic functions**.

5.2 Logic circuits

Aircraft logic systems follow the same conventions and standards as those used in other electronic applications. In particular, the MIL/ANSI standard logic symbols are invariably used and the logic elements that they represent operate in exactly the same way as those used in non-aircraft applications.

MIL/ANSI standard symbols for the most common logic gates are shown together with their truth tables in Figure 5.1.

5.2.1 Buffers

Buffers do not affect the logical state of a digital signal (i.e. a logic 1 input results in a logic 1 output whereas a logic 0 input results in a logic 0 output). Buffers are normally used to provide extra current drive at the output but can also be used to regularize the logic levels present at an interface.

5.2.2 Inverters

Inverters are used to complement the logical state (i.e. a logic 1 input results in a logic 0 output and vice versa). Inverters also provide extra current

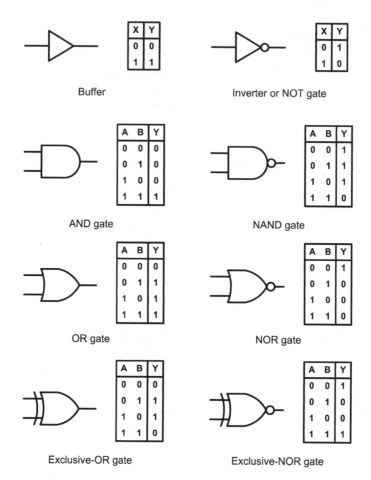

X	Y
0	0
1	1

Buffer

X	Y
0	1
1	0

Inverter or NOT gate

A	B	Y
0	0	0
0	1	0
1	0	0
1	1	1

AND gate

A	B	Y
0	0	1
0	1	1
1	0	1
1	1	0

NAND gate

A	B	Y
0	0	0
0	1	1
1	0	1
1	1	1

OR gate

A	B	Y
0	0	1
0	1	0
1	0	0
1	1	0

NOR gate

A	B	Y
0	0	0
0	1	1
1	0	1
1	1	0

Exclusive-OR gate

A	B	Y
0	0	1
0	1	0
1	0	0
1	1	1

Exclusive-NOR gate

Figure 5.1 MIL/ANSI symbols for standard logic gates together with truth tables

drive and, like buffers, are used in interfacing applications where they provide a means of regularizing logic levels present at the input or output of a digital system.

5.2.3 AND gates

AND gates will only produce a logic 1 output when all inputs are simultaneously at logic 1. Any other input combination results in a logic 0 output.

5.2.4 OR gates

OR gates will produce a logic 1 output whenever

any one, or more, inputs are at logic 1. Putting this another way, an OR gate will only produce a logic 0 output whenever all of its inputs are simultaneously at logic 0.

5.2.5 NAND gates

NAND (i.e. NOT-AND) gates will only produce a logic 0 output when all inputs are simultaneously at logic 1. Any other input combination will produce a logic 1 output. A NAND gate, therefore, is nothing more than an AND gate with its output inverted. The circle shown at the output of the gate denotes this inversion.

5.2.6 NOR gates

NOR (i.e. NOT-OR) gates will only produce a logic 1 output when all inputs are simultaneously at logic 0. Any other input combination will produce a logic 0 output. A NOR gate, therefore, is simply an OR gate with its output inverted. A circle is again used to indicate inversion.

5.2.7 Exclusive-OR gates

Exclusive-OR gates will produce a logic 1 output whenever either one of the two inputs is at logic 1 and the other is at logic 0. Exclusive-OR gates produce a logic 0 output whenever both inputs have the same logical state (i.e. when both are at logic 0 or both are at logic 1).

5.2.8 Exclusive-NOR gates

Exclusive-NOR gates will produce a logic 0 output whenever either one of the two inputs is at logic 1 and the other is at logic 0. Exclusive-OR gates produce a logic 1 output whenever both inputs have the same logical state (i.e. when both are at logic 0 or both are at logic 1).

5.2.9 Inverted outputs and inputs

The NAND and NOR gates that we have just met are said to have inverted outputs. In other words, they are respectively equivalent to AND and OR gates with their outputs passed through an inverter (or NOT gate) as shown in Figure 5.2(a) and Figure 5.2(b).

As well as inverted outputs, aircraft logic systems also tend to show logic gates in which one or more of the inputs are inverted. In Figure 5.2(c) an AND gate is shown with one input inverted. This is equivalent to an inverter (NOT gate) connected to one input of the AND gate, as shown. In Figure 5.2(d) an OR gate is shown with one input inverted. This is equivalent to an inverter (NOT gate) connected to one input of the OR gate, as shown.

Two further circuits with inverted inputs are shown in Figure 5.3. In Figure 5.3(a), both inputs of an AND gate are shown inverted. This

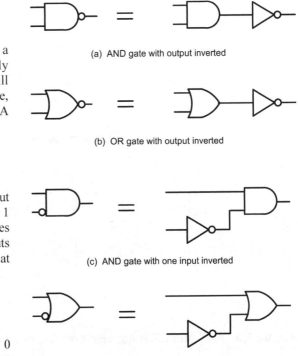

(a) AND gate with output inverted

(b) OR gate with output inverted

(c) AND gate with one input inverted

(d) OR gate with one input inverted

Figure 5.2 Gates with inverted outputs and inputs

arrangement is equivalent to the two-input NOR gate shown. In Figure 5.3(b), both inputs of an OR gate are shown inverted. This arrangement is equivalent to the two-input NAND gate shown.

5.3 Boolean algebra

Boolean algebra is frequently used to describe logical operations used in avionic systems. The basic Boolean logic expressions are:

$Y = A \cdot B$ (the AND function)

$Y = A + B$ (the OR function)

$Y = \overline{A \cdot B}$ (the NAND function)

$Y = \overline{A + B}$ (the NOR function)

$Y = \overline{A}$ (the NOT function)

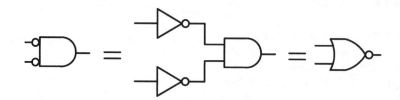

(a) AND gate with both inputs inverted

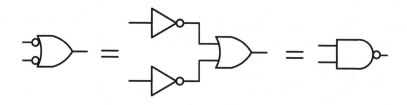

(b) OR gate with both inputs inverted

Figure 5.3 AND and OR gates with both inputs inverted

The rules (or laws) of Boolean algebra are as follows:

The Commutative Law

$A + B = B + A$

$A \cdot B = B \cdot A$

$A + (B + C) = (A + B) + C$

$A \cdot (B \cdot C) = (A \cdot B) \cdot C$

The Distributive Law

$A \cdot (B + C) = (A \cdot B) + (A \cdot C)$

The AND rules

$A \cdot 0 = 0$

$A \cdot 1 = A$

$A \cdot A = A$

$A \cdot \overline{A} = 0$

The OR rules

$A + 0 = A$

$A + 1 = 1$

$A + \overline{A} = 1$

$A + A = A$

The NOT rules

$\overline{0} = 1$

$\overline{1} = 0$

$\overline{\overline{A}} = A$ (double inversion)

Note that it is important to avoid confusing the symbols and laws of Boolean algebra with those that apply to conventional algebra! For example, in conventional algebra, $x + x = 2x$ whereas in Boolean algebra $X + X = X$!

De Morgan's Theorem

We have already shown how, when the two inputs to an AND gate are inverted the resulting logic is the same as that of a NOR gate. This follows from De Morgan's theorem and can be shown using Boolean algebra as:

$$\overline{A} \bullet \overline{B} = \overline{A + B}$$

Similarly, when the two inputs to an OR gate are inverted the resulting logic is the same as that of a NAND gate. Using Boolean algebra this is shown as:

$$\overline{A} + \overline{B} = \overline{A \bullet B}$$

It may help to remember the rule as 'split (or join) the bar and change the sign'.

In an avionic system the Boolean variables (A, B, C, etc) are usually replaced by the abbreviated names of signals, such as NOSE GEAR DOWN, IRS GND SPD >100 KTS, AUTOTRIM VALID, GROUND TEST REQUEST, etc. With a little practice, the use of Boolean algebra should become second nature!

A	B	C	Y
0	0	0	0
0	0	1	0
0	1	0	0
0	1	1	1
1	0	0	0
1	0	1	1
1	1	0	1
1	1	1	1

Figure 5.4 The majority vote truth table

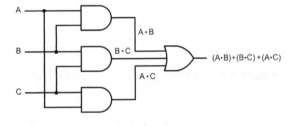
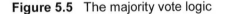

Figure 5.5 The majority vote logic

Key Point

Logic circuits involve signals that can only exist in one of two, mutually exclusive, states. These two states are usually denoted by 1 and 0, 'on' or 'off', 'high' and 'low', closed' and 'open', etc.

5.4 Combinational logic

By using a standard range of logic levels (i.e. voltage levels used to represent the logic 1 and logic 0 states) logic circuits can be combined together in order to solve more complex logic functions. As an example, assume that a logic circuit is to be constructed that will produce a logic 1 output whenever two, or more, of its three inputs are at logic 1. This circuit is referred to as a majority vote circuit and its truth table is shown in Figure 5.4. Figure 5.5 shows the logic circuitry required and the Boolean expressions for the logic at each node in the circuit.

Now let's look at a more practical example of the use of logic in the typical aircraft system shown in Figure 5.6. The inputs to this logic system consist of five switches that detect whether or not the respective landing gear door is open. The output from the logic system is used to

drive six warning indicators. Four of these are located on the overhead display panel and show which door (or doors) are left open whilst an indicator located on the pilot's instrument panel provides a master landing gear door warning. A switch is also provided in order to enable or disable the five door warning indicators.

The landing gear warning logic primary module consists of the following integrated circuit devices:

A1	Regulated power supply for A5
A2	Regulated power supply for A7 and A11
A5	Ten inverting (NOT) gates
A7	Five-input NAND gate
A11	Six inverting (NOT) gates

Note that the power supply for A1 and A2 is derived from the essential services DC bus. This is a 28 V DC bus which is maintained in the event of an aircraft generator failure. Note also that the indicators are active-low devices (in other words,

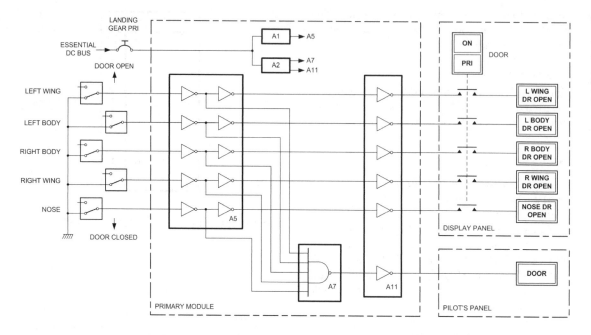

Figure 5.6 Landing gear door warning logic

they require a logic 0 input in order to become illuminated).

In order to understand how the landing gear warning logic works it is simply a matter of tracing logic 0 and logic 1 states through the logic diagram. Figure 5.7 shows how this is done when all of the landing gear doors are closed (this is the normal in-flight condition). Note how the primary door warning indicator shows the pilot that the system is active. When all of the landing gear doors are closed all inputs to A5 are taken to logic 0, all outputs from A5 are at logic 0 as is the output from A7. This, in turn, results in logic 1 inputs to the indicators which remain in the off (non-illuminated) state.

In Figure 5.8 the nose landing gear door is open. In this condition the output of A7 goes to logic 1 and the master warning becomes illuminated on the pilot's panel. At the same time, the nose door open warning becomes illuminated.

In Figure 5.9 both the left wing and the nose landing gear doors are open. In this condition the output of A7 goes to logic 1 and the master warning becomes illuminated as before. This time, however, both the nose door open and left wing door open warnings become illuminated.

Note that in a real passenger aircraft a secondary landing gear door warning logic system is fitted. This system is identical to the primary system shown and it provides a back-up in case the primary system fails. Primary or secondary system operation can be selected by the pilot.

Key Point

The logical function of a combinational logic circuit (i.e. an arrangement of logic gates) can be described by a truth table or by using Boolean algebra.

Test your understanding 5.1

1. Show how a four-input AND gate can be made using three two-input AND gates.
2. Show how a four-input OR gate can be made using three two-input OR gates.
3. Show how an exclusive-OR gate can be made by combining AND, OR and NOT gates.

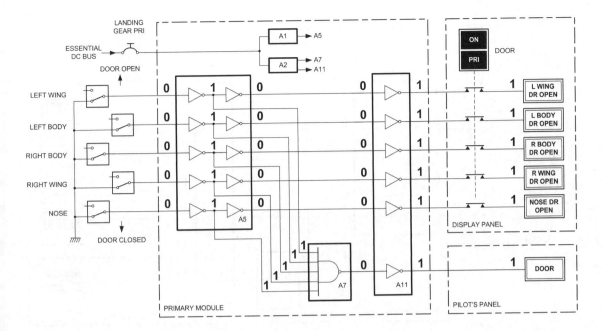

Figure 5.7 Landing gear door warning logic with all doors closed (normal in-flight condition)

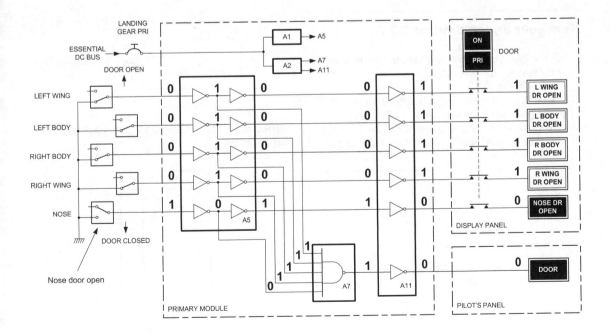

Figure 5.8 Landing gear door warning logic with nose door open

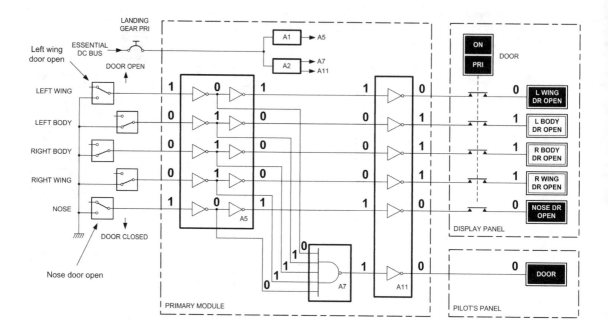

Figure 5.9　Landing gear door warning logic with nose and left wing door open

---

Test your understanding 5.2

1. Draw the truth table and state the Boolean expression for the logic gate arrangement shown in Fig. 5.10.
2. State the Boolean logic expression for the logic gate arrangement shown in Fig. 5.11.
3. Devise a logic gate arrangement that provides the output described by the truth table shown in Fig. 5.12.

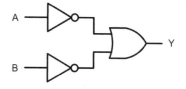

Figure 5.11　See Question 2 of Test your understanding 5.2

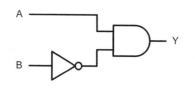

Figure 5.10　See Question 1 of Test your understanding 5.2

A	B	C	Y
0	0	0	0
0	0	1	0
0	1	0	1
0	1	1	1
1	0	0	0
1	0	1	1
1	1	0	1
1	1	1	1

Figure 5.12　See Question 3 of Test your understanding 5.2

5.5 Tri-state logic

Tri-state logic devices operate in a similar manner to conventional logic gates but have a third, high-impedance output state. This high-impedance state permits the output of several tri-state devices to be connected directly together. Such arrangements are commonly used when a bus system has to be driven by several logic gates. The output state (whether a logic level or whether high-impedance) is controlled by means of an enable (EN) input. This EN input may be either active-high or active-low, as shown in Fig. 5.13.

Figures 5.13(a) and 5.13(b) show the truth table for a buffer with active-high and active-low enable inputs respectively. Note that the state of the EN input determines whether the output takes the same logical state as the input or whether the output is taken to its high-impedance state (shown as X in the truth table).

Figures 5.13(c) and 5.13(d) show the truth table for an inverter with active-high and active-low enable inputs respectively. In this case, the state of the EN input determines whether the output takes the opposite logical state to the input or whether the output is taken to its high-impedance state (again shown as X in the truth table).

5.6 Monostables

Monostable (or **one-shot**) devices provide us with a means of generating precise time delays. Such delays become important in many **sequential logic** applications where logic states are not constant but subject to change with time.

The action of a monostable is quite simple—its output is initially logic 0 until a change of state occurs at its **trigger input**. The level change can be from 0 to 1 (positive edge trigger) or 1 to 0 (negative edge trigger). Immediately the trigger pulse arrives, the output of the monostable changes state to logic 1. It then remains at logic 1 for a pre-determined period before reverting back to logic 0. Monostable circuits can be used as pulse stretchers (a means of elongating a short pulse) as well as producing accurate time delays.

An example of the use of a monostable is shown in the Auxiliary Power Unit (APU) starter

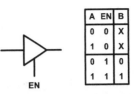

A	EN	B
0	0	X
1	0	X
0	1	0
1	1	1

(a) Tri-state buffer with active-high enable

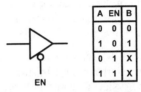

A	EN	B
0	0	0
1	0	1
0	1	X
1	1	X

(b) Tri-state buffer with active-low enable

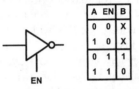

A	EN	B
0	0	X
1	0	X
0	1	1
1	1	0

(c) Tri-state inverter with active-high enable

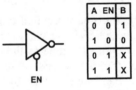

A	EN	B
0	0	1
1	0	0
0	1	X
1	1	X

(d) Tri-state inverter with active-low enable

Figure 5.13 Tri-state logic devices

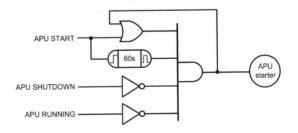

Figure 5.14 APU starter logic

logic shown in Figure 5.14. This arrangement has three inputs (APU START, APU SHUTDOWN, and APU RUNNING) and one output (APU STARTER MOTOR). The inputs are all active-high (in other words, a logic 1 is generated when the pilot operates the APU START switch, and so on). The output of the APU starter motor control logic goes to logic 1 in order to apply power to the starter motor via a large relay.

There are a few things to note about the logic arrangement shown in Figure 5.14:

1. When the APU runs on its own we need to disengage the starter motor. In this condition the APU MOTOR signal needs to become inactive (i.e. it needs to revert to logic 0).

2. We need to avoid the situation that might occur if the APU does not start but the starter motor runs continuously (as this will drain the aircraft batteries). Instead, we should run the starter motor for a reasonable time (say, 60 seconds) before disengaging the starter motor. The 60 seconds timing is provided by means of a positive edge triggered monostable device. This device is triggered from the APU START signal.

3. Since the pilot is only required to momentarily press the APU START switch, we need to hold the condition until such time as the engine starts or times out (i.e. at the end of the 60 second period). We can achieve this by OR'ing the momentary APU START signal with the APU STARTER MOTOR signal.

4. We need to provide a signal that the pilot can use to shutdown the APU (for example, when the aircraft's main engines are running or perhaps in the event of a fault condition).

In order to understand the operation of the APU starter motor logic system we can once again trace through the logic system using 1s and 0s to represent the logical condition at each point (just as we did for the landing gear door warning logic).

In Figure 5.15(a) the APU is in normal flight and the APU is not running. In this condition the main engines are providing the aircraft's electrical power.

In Figure 5.15(b) the pilot is operating the APU START switch. The monostable is triggered and output of the OR and AND gates both go to

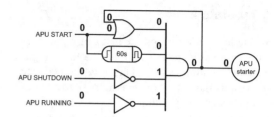

(a) Normal flight; engine power generation; APU not running

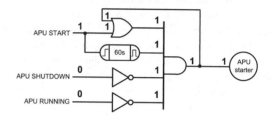

(b) APU starter switch operated; APU starter motor begins to run

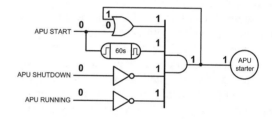

(c) APU starter motor continues to run for up to 60s

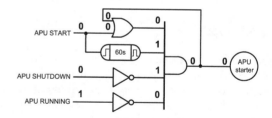

(d) APU runs before 60s timeout; starter motor stops when APU runs

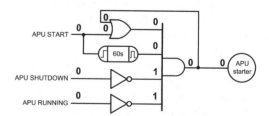

(e) APU fails to run during 60s period; further APU START signal awaited

Figure 5.15 APU starter operation

logic 1 in order to assert the APU STARTER MOTOR signal.

In Figure 5.15(c) the APU START signal is removed but the output of the AND gate is held at logic 1 by feeding back its logical state via the OR gate. The monostable remains triggered and continues to produce a logic 1 output for its 60 second period.

In Figure 5.15(d) the APU is now running and the APU RUNNING signal has gone to logic 1 in order to signal this conditions. This results in the output of the AND gate going to logic 0 and the APU STARTER MOTOR signal is no longer made active. The starter motor is therefore disengaged.

In Figure 5.15(e) the APU has failed to run during the 60 second monostable period. In this **timed out** condition the output of the AND gate goes to logic 0 and the APU STARTER MOTOR signal becomes inactive. The system then waits for the pilot to operate the APU START button for a further attempt at starting!

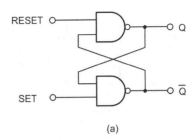

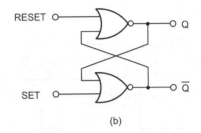

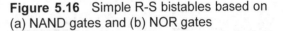

Figure 5.16 Simple R-S bistables based on (a) NAND gates and (b) NOR gates

5.7 Bistables

The output of a bistable circuit has two stable states (logic 0 or logic 1). Once **set** in one or other of these states, the output of a bistable will remain at a particular logic level for an indefinite period until **reset**. A bistable thus forms a simple form of memory as it remains in its latched state (either **set** or **reset**) until a signal is applied to it in order to change its state (or until the supply is disconnected).

The simplest form of bistable is the R-S bistable. This device has two inputs, SET and RESET, and complementary outputs, Q and /Q. A logic 1 applied to the SET input will cause the Q output to become (or remain at) logic 1 while a logic 1 applied to the RESET input will cause the Q output to become (or remain at) logic 0. In either case, the bistable will remain in its SET or RESET state until an input is applied in such a sense as to change the state.

Two simple forms of R-S bistable based on cross-coupled logic gates are shown in Figure 5.16. Figure 5.16(a) is based on cross-coupled two-input NAND gates while Figure 5.16(b) is based on cross-coupled two-input NOR gates.

Unfortunately, the simple cross-coupled logic gate bistable has a number of serious shortcomings (consider what would happen if a logic 1 was simultaneously present on both the SET and RESET inputs!) and practical forms of bistable make use of much improved purpose-designed logic circuits such as D-type and J-K bistables.

The D-type bistable has two inputs: D (standing variously for 'data' or 'delay') and CLOCK (CLK). The data input (logic 0 or logic 1) is clocked into the bistable such that the output state only changes when the clock changes state. Operation is thus said to be synchronous. Additional subsidiary inputs (which are invariably active low) are provided which can be used to directly set or reset the bistable. These are usually called PRESET (PR) and CLEAR (CLR). D-type bistables are used both as latches (a simple form of memory) and as binary dividers. The simple circuit arrangement in Figure 5.17 together with the **timing diagram** shown in Figure 5.18 illustrate the operation of D-type bistables.

J-K bistables (see Figure 5.19) have two clocked inputs (J and K), two direct inputs

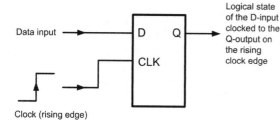

Figure 5.17 D-type bistable

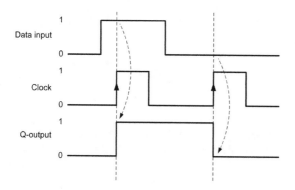

Figure 5.18   Timing diagram for the D-type bistable

(PRESET and CLEAR), a CLOCK (CLK) input, and outputs (Q and /Q). As with R-S bistables, the two outputs are complementary (i.e. when one is 0 the other is 1, and vice versa). Similarly, the PRESET and CLEAR inputs are invariably both active low (i.e. a 0 on the PRESET input will set the Q output to 1 whereas a 0 on the CLEAR input will set the Q output to 0). Figure 5.20 summarizes the input and corresponding output states of a J-K bistable for various input states. J-K bistables are the most sophisticated and flexible of the bistable types and they can be configured in various ways for use in binary dividers, shift registers, and latches.

5.7.1 Binary counters

Figure 5.21 shows the arrangement of a four-stage binary counter based on J-K bistables. The timing diagram for this circuit is shown in Figure 5.22. Each stage successively divides the clock

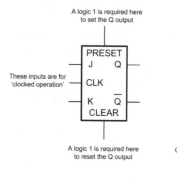
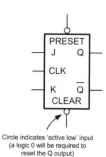

(a) Basic J-K bistable

(b) J-K bistable with active-low PRESET and CLEAR inputs

Figure 5.19   J-K bistable symbols

(a) PRESET and CLEAR inputs

Inputs		Output	Comment
PRESET	CLEAR	Q	
0	0	?	Indeterminate
0	1	0	Q output changes to 0 (i.e. Q is reset) regardless of the clock
1	0	1	Q output changes to 1 (i.e. Q is reset) on the next clock transition)
1	1	-	Enables clocked operation - refer to the next truth table

Note that the PRESET and CLEAR inputs are unaffected by the state of the clock

(b) Clocked operation using the J and K inputs

Inputs		Output	Comment
J	K	Q_{N+1}	
0	0	Q_N	No change in state of the Q output on the next clock transition
0	1	0	Q output changes to 0 (i.e. Q is reset) on the next clock transition)
1	0	1	Q output changes to 1 (i.e. Q is reset) on the next clock transition)
1	1	Q_N	Q output changes to the opposite state on the next clock transition

Note that Q_N means 'Q in whatever state it was before' whilst Q_{N+1} means 'Q after the next clock transition'

Figure 5.20   Truth tables for the J-K bistable

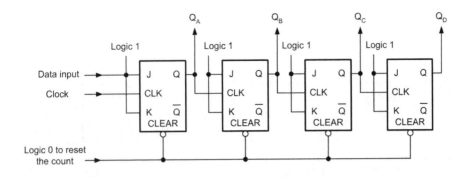

Figure 5.21 Four-stage binary counter using J-K bistables

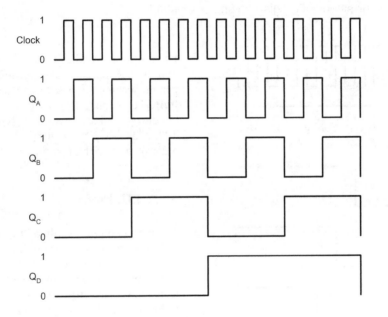

Figure 5.22 Timing diagram for the four-stage binary counter

input signal by a factor of two. Note that a logic 1 input is transferred to the respective Q-output on the falling edge of the clock pulse and all J and K inputs must be taken to logic 1 to enable binary counting.

5.7.2 Shift registers

Figure 5.23 shows the arrangement of a four-stage shift register based on J-K bistables. The timing diagram for this circuit is shown in Figure

5.24. Note that each stage successively feeds data (via the Q output) to the next stage and that all data transfer occurs on the falling edge of the clock pulse.

5.8 Logic families

The task of realizing a complex logic circuit is made simple with the aid of digital integrated circuits which are classified according to the semiconductor technology used in their

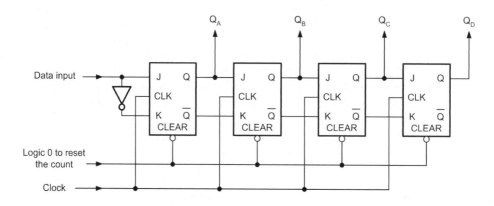

Figure 5.23 Four-stage shift register using J-K bistables

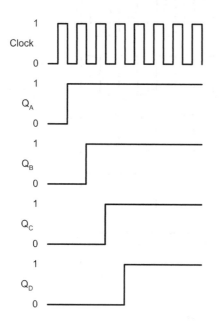

Figure 5.24 Timing diagram for the four-stage shift register

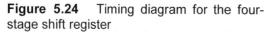

fabrication (the logic family to which a device belongs is largely instrumental in determining its operational characteristics, such as power consumption, speed, and immunity to noise, see Table 5.1).

The two basic logic families are complementary metal oxide semiconductor (CMOS) and transistor-transistor logic (TTL).

Each of these families is then further sub-divided into classes that are based on refinements of the parent technology, such as 'high-current' (or **buffered** output), low-noise, etc. Representative circuits for a basic two-input NAND gate using TTL and CMOS technology are shown in Figures 5.25(a) and 5.25(b), respectively.

5.8.1 TTL logic

The most common family of TTL logic devices is the 74-series. Devices from this family are coded with the prefix number 74, for example, 7400. Sub-families (based on the use of variations in the technology) are distinguished by letters which follow the initial prefix, for example:

74F00   A logic device that uses fast TTL technology and operates at higher speed than the 7400 with which it is logic and pin-compatible

74LS08   A logic device that is logic and pin-compatible with the 7408 but which uses low-power Schottky technology

74HCT74   A high-speed CMOS version of the 7474 logic device which has CMOS-compatible inputs

74ALS132   A logic device that is logic and pin-compatible with the 7414 but which uses advanced low-power Schottky technology.

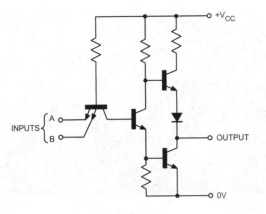

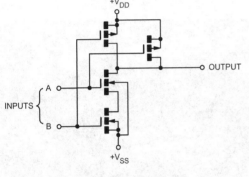

(a) Standard TTL logic

(b) Standard CMOS logic

Figure 5.25 Representative internal circuitry for a two-input NAND gate in (a) standard TTL and (b) standard CMOS technology

5.8.2 CMOS logic

The most common family of CMOS devices is the 4000-series. Variants within the family are identified by the suffix letters, for example:

4001A	A standard (un-buffered) CMOS logic device that is logic and pin-compatible with the 4001
4011B	A logic device that is logic and pin-compatible with the 4001 but which has buffered outputs
4069UBE	A logic device that is logic and pin-compatible with the 4069 but which has un-buffered outputs.

5.8.3 Logic levels and noise margin

Logic levels are simply the range of voltages used to represent the logic states 0 and 1. The logic levels for CMOS differ markedly from those associated with TTL. In particular, CMOS logic levels are relative to the supply voltage used while the logic levels associated with TTL devices tend to be absolute (see Fig. 5.26).

The **noise margin** of a logic device is a measure of its ability to reject noise and spurious signals; the larger the noise margin the better is its ability to perform in an environment in which noise is present. Noise margin is defined as the difference between the minimum values of high state output and high state input voltage and the

Table 5.1 Comparison of typical performance specifications for various logic families

Characteristic	Logic family			
	Std. TTL	Std. CMOS	LS-TTL	HC-TTL
Maximum recommended supply voltage (V)	5.25	18	5.25	5.5
Minimum recommended supply voltage (V)	4.75	3	4.75	4.5
Static power dissipation (mW per gate at 100 kHz)	10	negligible	2	negligible
Typical propagation delay (ns)	10	105	10	10
Maximum clock frequency (MHz)	35	12	40	40
Minimum output current (mA)	16	1.6	8	4
Fan-out (maximum number of standard loads)	40	4	20	10

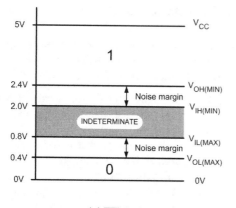

(a) TTL logic

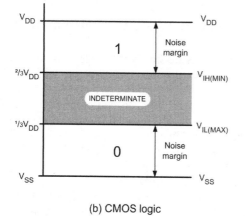

(b) CMOS logic

Figure 5.26 Logic levels for (a) standard TTL and (b) standard CMOS logic devices

Figure 5.27 Logic circuits mounted on a microcontroller printed circuit board

Figure 5.28 A low-power Schottky TTL logic circuit supplied in a 14-pin DIL package

maximum values of low state output and low state input voltage. Hence:

Noise margin = $V_{OH(MIN)} - V_{IH(MIN)}$, and also

Noise margin = $V_{OL(MAX)} - V_{IL(MAX)}$

Where $V_{OH(MIN)}$ is the minimum value of high state (logic 1) output voltage, $V_{IH(MIN)}$ is the minimum value of high state (logic 1) input voltage, $V_{OL(MAX)}$ is the maximum value of low state (logic 0) output voltage, and $V_{IL(MAX)}$ is the minimum value of low state (logic 0) input voltage.

The noise margin for standard 7400 series TTL is typically 400 mV while that for CMOS is (1/3) V_{DD}, as shown in Figure 5.26.

Key Point

The two major logic families are CMOS and TTL. These two families have quite different characteristics, supply requirements, and logic levels. CMOS logic devices operate from a wide range of voltage levels and consume very little power (and are therefore often preferred for portable low-power applications). TTL logic devices, on the other hand, tend to be faster but have a much lower noise margin and are therefore more susceptible to noise and interference. The two main logic families are further divided into a number of different sub-families based on variations in the parent technology.

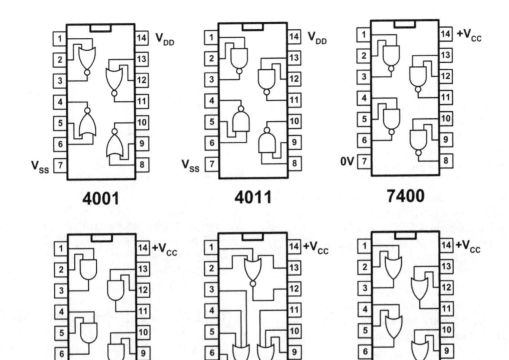

Figure 5.29 Pin connections for some common CMOS and TTL devices supplied in 14-pin dual-in-line (DIL) packages (see Test your understanding 5.3)

Test your understanding 5.3

Fig. 5.29 shows the pin connections for some common CMOS and TTL logic devices.

1. Which of the devices shown is a two-input OR gate?

2. What voltage would you expect to measure on pin 14 of a 7408 device?

2. Sketch a circuit (including pin numbers) showing how a 4001 device could be used as a dual R-S bistable.

3. Sketch circuits (including pin numbers) showing how a 7400 could be used as: (a) a two-input AND gate and (b) a two-input OR gate.

5.9 Multiple choice questions

1. The logic device shown in Figure 5.30 is:
   (a) an OR gate
   (b) a NOR gate.
   (c) an exclusive-OR gate.

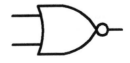

Figure 5.30 See Question 1.

2. The normal supply voltage for a TTL logic device is:
   (a) 2.5 V ±5%
   (b) 5 V ±5%
   (c) 12 V ±5%.

3. A two-input NAND gate will produce a logic 0 output when:
   (a) both of the inputs are at logic 0
   (b) either one of the inputs is at logic 0
   (c) both of the inputs are at logic 1.

4. In a binary counter, the clock input of each bistable stage is fed from:
   (a) the same clock line
   (b) the Q output of the previous stage
   (c) the CLEAR line.

5. The device shown in Figure 5.31 is:
   (a) a low-power Schottky TTL device
   (b) a high density standard TTL device
   (c) a buffered CMOS device.

Figure 5.31   See Question 5.

6. The most appropriate logic family for use in a portable item of test equipment is:
   (a) CMOS
   (b) TTL
   (c) Low-power Schottky TTL.

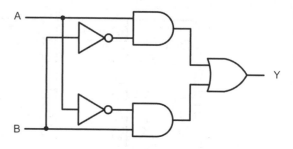

Figure 5.32   See Question 7.

7. The logic gate arrangement shown in Figure 5.32 performs the same function as:
   (a) a NOR gate
   (b) a NAND gate
   (c) an exclusive-OR gate.

8. The noise margin for standard TTL devices is:
   (a) 400 mV
   (b) 800 mV
   (c) 2 V.

9. A CMOS logic gate is operated from a 12 V supply. If a voltage of 3 V is measured at the input to the gate, this would be considered equivalent to:
   (a) logic 0
   (b) indeterminate
   (c) logic 1.

10. The truth table shown in Figure 5.33 is for:

    (a) a two-input OR gate
    (b) a two input NOR gate

A	B	Y
0	0	1
0	1	0
1	0	0
1	1	0

Figure 5.33   See Question 10.

Chapter 6 Computers

Modern aircraft use increasingly sophisticated avionic systems which involve the use of microprocessor-based computer systems. These systems combine hardware and software and are capable of processing large amounts of data in a very small time. This chapter provides an overview of aircraft computer systems.

6.1 Computer systems

The basic components of a computer system are shown in Figure 6.1. The main components are:

(a) a **central processing unit (CPU)**
(b) a **memory**, comprising both 'read/write' and 'read only' devices (commonly called **RAM** and **ROM** respectively)
(c) a means of providing **input** and **output (I/O)**. For example, a keypad for input and a display for output.

In a microprocessor system the functions of the CPU are provided by a single very large scale integrated (VLSI) microprocessor chip (see Chapter 7). This chip is equivalent to many thousands of individual transistors.

Semiconductor devices are also used to provide the read/write and read-only memory. Strictly speaking, both types of memory permit 'random access' since any item of data can be retrieved with equal ease regardless of its actual location within the memory. Despite this, the term 'RAM' has become synonymous with semiconductor read/write memory.

The basic components of the system (CPU, RAM, ROM and I/O) are linked together using a multiple-wire connecting system know as a **bus** (see Figure 6.1). Three different buses are present, these are:

(a) the **address bus** used to specify memory locations;
(b) the **data bus** on which data is transferred between devices; and
(c) the **control bus** which provides timing and control signals throughout the system.

The number of individual lines present within the address bus and data bus depends upon the particular microprocessor employed. Signals on all lines, no matter whether they are used for address, data, or control, can exist in only two basic states: logic 0 (**low**) or logic 1 (**high**). Data and addresses are represented by **binary numbers** (a sequence of 1s and 0s) that appear respectively on the data and address bus.

Some basic microprocessors designed for control and instrumentation applications have an 8-bit data bus and a 16-bit address bus. More

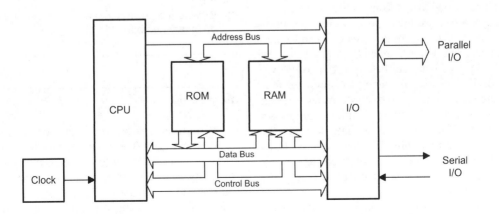

Figure 6.1 Basic components of a computer system

sophisticated processors can operate with as many as 64 or 128 bits at a time.

The largest binary number that can appear on an 8-bit data bus corresponds to the condition when all eight lines are at logic 1. Therefore the largest value of data that can be present on the bus at any instant of time is equivalent to the binary number 11111111 (or 255). Similarly, the highest most address that can appear on a 16-bit address bus is 1111111111111111 (or 65,535). The full range of data values and addresses for a simple microprocessor of this type is thus:

Data	from	00000000
	to	11111111
Addresses	from	0000000000000000
	to	1111111111111111

Finally, a locally generated clock signal provides a time reference for synchronizing the transfer of data within the system. The clock usually consists of a high-frequency square wave pulse train derived from a quartz crystal (see Chapter 8).

6.2 Data representation

As we have learned from Chapter 2, binary numbers—particularly large ones—are not very convenient. To make numbers easier to handle we introduced the **hexadecimal** (base 16) numbering system. This format is easier for mere humans to comprehend and also offers the significant advantage over base 10 numbers in that numbers can be converted to and from binary with ease.

A group of eight bits, operated on as a unit, is referred to as a **byte**. Since hexadecimal characters can be represented by a group of four bits, a byte of data can be expressed using two hexadecimal characters. Note that groups of four bits are sometimes referred to as **nibbles** and that a byte can take a hexadecimal value ranging from 00 to FF. The basic unit of data that can be manipulated as an entity is often referred to as **word**. Words can be any convenient length but 16, 32 and 64-bit words are common (see Table 6.1).

A single byte of data can be stored at each address within the total memory space of a computer system. Hence one byte can be stored at

each of the 65,536 memory locations within a microprocessor system having a 16-bit address bus. Individual bits within a byte are numbered from 0 (least significant bit, or **LSB**) to 7 (most significant bit, **MSB**). In the case of 16-bit words (which are stored in consecutive memory locations) the bits are numbered from 0 (LSB) to 15 (MSB).

Negative numbers (or **signed numbers**) are usually be represented using **two's complement** (see page 18) notation where the leading (most significant) bit indicates the sign of the number (1 = negative, 0 = positive). For example, the signed 8-bit number 10000001 represents the decimal number −1.

Table 6.1   Some common data types

Data type	Bits	Range of values
Unsigned byte	8	0 to 255
Signed byte	8	−128 to +127
Unsigned word	16	0 to 65,535
Signed word	16	−32,768 to +32,767
Unsigned double word	32	0 to 4,294,967,296

Test your understanding 6.1

1. What is the highest data value (expressed in decimal) that can appear on a 12-bit address bus?
2. What is the decimal equivalent of the signed 8-bit binary number 11111110?

Key Point

A computer system consists of a central processing unit (CPU), a read-only memory (ROM), a read/write (random access) memory (RAM), and one or more input/output (I/O) devices. These elements are linked together using a local bus system that comprises an address bus, a data bus, and a control bus.

6.3 Data storage

The semiconductor ROM within a microprocessor system provides storage for the program code as well as any permanent data that requires storage. All of this data is referred to as **non-volatile** because it remains intact when the power supply is disconnected.

The semiconductor RAM within a microprocessor system provides storage for the transient data and variables that are used by programs. Part of the RAM is also used by the microprocessor as a temporary store for data whilst carrying out its normal processing tasks.

It is important to note that any program or data stored in RAM will be lost when the power supply is switched off or disconnected. The only exception to this is low-power CMOS RAM that is kept alive by means of a small battery. This **battery-backed memory** is used to retain important data, such as the time and date.

When expressing the amount of storage provided by a memory device we usually use Kilobytes (Kbyte). It is important to note that a Kilobyte of memory is actually 1,024 bytes (not 1,000 bytes). The reason for choosing the Kbyte rather than the kbyte (1,000 bytes) is that 1,024 happens to be the nearest power of 2 (note that $2^{10} = 1,024$).

The capacity of a semiconductor ROM is usually specified in terms of an address range and the number of bits stored at each address. For example, 2 K × 8 bits (capacity 2 Kbytes), 4 K × 8 bits (capacity 4 Kbytes), and so on. Note that it is not always necessary (or desirable) for the entire memory space of a computer to be populated by memory devices.

6.3.1 Random access memory

A large proportion (typically 80% or more) of the total addressable memory space of a computer system is devoted to read/write memory. This area of memory is used for a variety of purposes, the most obvious of which is program and data storage. The term 'random access' simply refers to a memory device in which data may be retrieved from all locations with equal ease (i.e.

access time is independent of actual memory address). This is important since our programs often involve moving sizeable blocks of data into and out of memory.

The basic element of a semiconductor memory is known as a **cell**. Cells can be fabricated in one of two semiconductor technologies: MOS (metal oxide semiconductor) and bipolar. Bipolar memories are now rarely used even though they offer much faster access times. Their disadvantage is associated with power supply requirements (they need several voltage rails, both positive and negative, and use significantly more power than their MOS counterparts).

Random access memories can be further divided into **static** and **dynamic** types. The important difference between the two types is that dynamic memories require periodic refreshing if they are not to lose their contents. While, in the normal course of events, this would be carried out whenever data was read and rewritten, this technique cannot be relied upon to refresh all of the dynamic memory space and steps must be taken to ensure that all dynamic memory cells are refreshed periodically. This function has to be integrated with the normal operation of the computer system or performed by a dedicated dynamic memory controller. Static memories do not need refreshing and can be relied upon to retain their memory until such time as new data is written or the power supply is interrupted (in which case all data is lost).

The circuit of a typical bipolar static memory cell is shown in Figure 6.2. The transistors form an R-S bistable (see page 55) that can be SET or RESET by means of a pulse applied to the appropriate emitter. The CELL SELECT line is used to identify the particular cell concerned. This type of cell uses emitter-coupled logic (ECL) and requires both negative and positive supply rails.

The circuit of a typical MOS static memory cell is shown in Figure 6.3. This is also clearly recognizable as a bistable element. The CELL SELECT line is used to gate signals into and out of the memory cell. Note that, as for the bipolar cell, the SET and RESET lines are common to a number of cells.

The circuit of a typical MOS dynamic memory cell is shown in Figure 6.4. In contrast to the

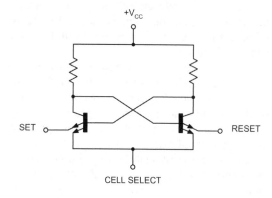

Figure 6.2 Circuit diagram of a typical bipolar static memory cell

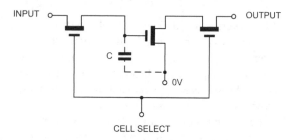

Figure 6.4 Circuit diagram of a typical MOS dynamic memory cell

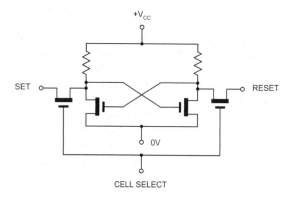

Figure 6.3 Circuit diagram of a typical MOS static memory cell

previous types of cell this clearly does not use a bistable arrangement. Instead, a capacitor, C, is used as the storage element. The capacitor is in fact the input capacitance of an MOS transistor. This is charged or discharged according to the state of the INPUT, OUTPUT and CELL SELECT lines.

The characteristics of a number of common random access memories are shown in Table 6.2. Both static (**SRAM**) and dynamic (**DRAM**) examples have been included. Note that there is some variation in the way that these memories are organized.

From Table 6.2 it should readily be apparent that individual random access memories must contain some form of internal decoding in order to make each cell individually addressable. This is achieved by arranging the cells in the form of a matrix. Figure 6.5 shows one possible arrangement where 16,384 individual memory

Table 6.2 Examples of various type of RAM

Type	Size (bits)	Organization	Package	Technology	Supply
4164	64K	64K words × 1 bit	16-pin DIL	NMOS DRAM	+5 V at 35 mA max.
424256	1M	256K words × 4 bit	20-pin DIL	CMOS DRAM	+5 V at 70 mA max.
62256	256K	32K words × 8 bit	28-pin DIL	CMOS DRAM	+5 V at 25 mA max.
6264	64K	8K words × 8 bit	28-pin DIL	CMOS SRAM	+5 V at 110 mA max.
DS1230AB	256K	32K words × 8 bit	28-pin DIL	CMOS SRAM	+5 V at 85 mA max.
TC514400	4M	1M words × 4 bit	26-pin TSOP	CMOS DRAM	+5 V at 100 mA max.

cells form a matrix consisting of 128 rows × 128 columns. Each cell within the array has a unique address and is selected by placing appropriate logic signals on the row and column address lines. All that is necessary to interface the memory matrix to the system is some additional logic, but first we must consider the mechanism which allows the address bus to be connected to the memory cell.

The row and column decoders of Figure 6.5 have 128 output lines and seven input lines. Each possible combination of the seven input lines results in the selection of a unique output line. If we assume that the row decoder handles the most significant part of the address while the column decoder operates on the least significant portion, the cell at the top left-hand corner of the matrix will correspond to memory location 0 (memory address 0000h) whereas the corresponding position in the next row will be 128 (memory address 0080h). Finally, the cell at the bottom right-hand corner will be 16,383 (memory address 3FFFh).

While it would be possible to have 14 separate address lines fed into the chip this is somewhat inelegant since, with the aid of some additional gating and latches, it is possible to reduce the number of address lines to seven. This is achieved by multiplexing part of the address bus and then de-multiplexing it within the memory device. A typical arrangement is shown in Figure 6.6. Separate row and column address select (**RAS** and **CAS**) control signals are required in order to enable the appropriate latches when the multiplexed address information is valid. The multiplexing arrangement for the address bus is shown in Figure 6.7. Note that, in practice, additional control signals will be required when several 16K blocks of RAM are present within the system. Such a precaution is essential in order to prevent data conflicts in which more than one memory cell (in different RAM blocks) is addressed simultaneously.

Table 6.3 shows the truth table for the address lines and decoding logic of the memory device. This table assumes that the most significant part

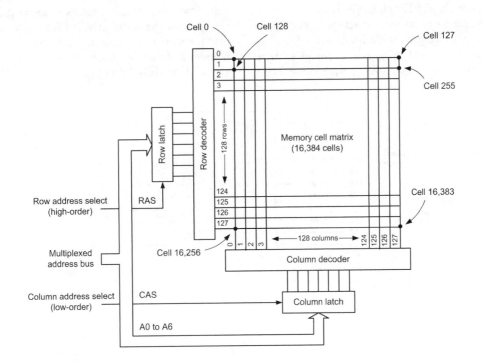

Figure 6.5 Possible arrangement for a 128 x 128 memory cell matrix

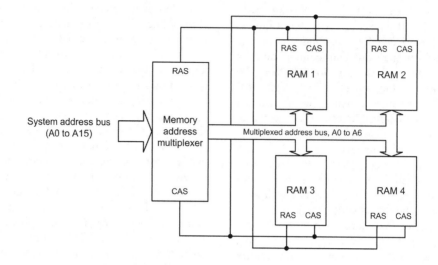

Figure 6.6 Basic arrangement for multiplexing the address bus

of the address (high order) appears first and is decoded to the appropriate row while the least significant (low order) part appears second and is decoded to the appropriate column. The table shows the logic states for three address locations: 0000h, 00FFh, and 3FFFh respectively.

The memory matrix described earlier is capable of storing a single bit at any of 16,384 locations (16K × 1 bit). For a complete byte (eight bits) we would obviously require eight

such devices. These would share the same address, RAS and CAS lines but each chip would be responsible for different bits of the data bus. Various methods are employed for data transfer into and out of the matrix, sensing the state of either the rows or the columns of the matrix and gating with the write enable (**WE**) line so that data transfers only occur at the correct time. Data output is generally tri-state (see page 53) by virtue of the shared data bus.

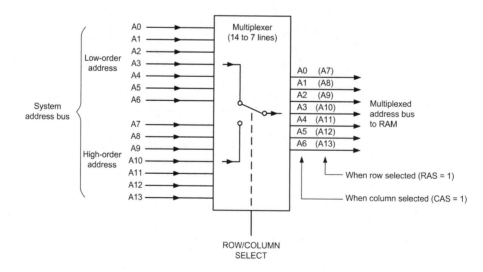

Figure 6.7 Multiplexer arrangement

Table 6.3 Multiplexed address decoding for a typical memory cell matrix

Order	A6	A5	A4	A3	A2	A1	A0	RAS	CAS	Row	Col.	Cell no.
High	0	0	0	0	0	0	0	1	0	0		0
Low	0	0	0	0	0	0	0	0	1		0	
High	0	0	0	0	0	0	1	1	0	1		255
Low	1	1	1	1	1	1	1	0	1		127	
High	1	1	1	1	1	1	1	1	0	127		16,383
Low	1	1	1	1	1	1	1	0	1		127	

6.3.2 Read only memory

As its name implies, read only memory is memory which, once programmed, can only be read from and not written to. It may thus be described as 'non-volatile' since its contents are not lost when the supply is disconnected. This facility is, of course, necessary for the long-term semi-permanent storage of operating systems and high-level language interpreters. To change the control program or constant data it is necessary to replace the ROM. This is a simple matter because ROMs are usually plug-in devices or can be programmed electrically. The following types are in common use:

(a) **Mask programmed ROM**. This, relatively expensive, process is suitable for very high volume production (several thousand units, or more) and involves the use of a mask that programs links within the ROM chip. These links establish a permanent pattern of bits in the row/column matrix of the memory. The customer (computer manufacturer) must supply the ROM manufacturer with the programming information from which the mask is generated.

(b) **One-time programmable electrically programmable ROM (OTP EPROM)**. This is a somewhat less expensive process than mask programming and is suitable for small/medium scale production. The memory cells consist of nichrome or polysilicon fuse links between rows and columns. These links can, by application of a suitable current pulse, be open-circuited or 'blown'. OTP PROMs are ideal for

Figure 6.8 RAM (left) and UV EPROM (right) used in a small computer system

prototype use and programming can be carried out by the computer manufacturer (or the component supplier) using relatively inexpensive equipment. When an OTP PROM has been thoroughly tested (and provided that high volume production can be envisaged) it is possible for the device to be replaced by a conventional mask programmed ROM.

(c) **Erasable PROM (UV EPROM)**. Unlike the two previous types of ROM, the EPROM can be re-programmed. EPROMs are manufactured with a window that allows light to fall upon the semiconductor memory cell matrix. The EPROM may be erased by exposure to a strong ultraviolet light source over a period of several minutes, or tens of minutes. Once erasure has taken place any previously applied bit pattern is completely removed, the EPROM is 'blank' and ready for programming. The programming process is carried out by the manufacturer from master software using a dedicated programming device which supplies pulses to establish the state of individual memory cells. This process usually takes several minutes (though some EPROM programmers can program several devices at once) and, since EPROMs tend to be relatively expensive, this process is clearly unsuitable for anything other than very small scale production. Furthermore, it should be noted that EPROMs tend to have rather different characteristics from PROMs and ROMs and thus subsequent volume production replacement may cause problems.

(d) **Electrically erasable PROM (EEPROM)** (also known as **Flash memory**). This type of ROM, can both be read to and written from. However, unlike RAM, Flash ROM is unsuitable for use in the read/write memory section of a computer since the writing process takes a considerable amount of time (typically a thousand times longer than the reading time). EEPROMs are relatively recent and relatively expensive devices. It's also worth noting that, until recently, a reasonable compromise for semi-permanent data and program storage could take the form of low-power consumption RAM fitted with back-up batteries (in certain circumstances such a system can be relied upon to retain stored information for a year, or more).

The internal matrix structure of read only memories is somewhat similar to that employed for random access memories, i.e. individual cells within the matrix are uniquely referenced by means of row and column address lines. ROMs vary in capacity and 64K, 128K and 256K devices are commonplace. The characteristics of some common read only memories are shown in Table 6.4.

Finally, it is important to note that ROM and EPROM devices do not lose their data when the supply is disconnected. They are thus said to be **non-volatile memories (NVM)**. Most RAM devices, unless battery-backed (see page 65), are **volatile memories** and the stored data will be lost when the supply is disconnected.

Test your understanding 6.2

What is the total capacity, expressed in Kb, of a memory that is organised as 32K words × 4 bits?

Key Point

RAM and ROM devices provide storage for data and programs within a computer system. Various semiconductor technologies are used to produce RAM and ROM devices and this impacts on the characteristics, performance and ultimate application for a particular memory device. Computer memories are either volatile or non-volatile (the latter preserving its contents when the power supply is disconnected). ROM and EPROM devices are non-volatile whilst most types of RAM are volatile, unless battery-backed.

Table 6.4　Examples of various types of ROM

Type	Size (bits)	Organization	Package	Technology	Supply
2764	64K	64K words × 8 bit	28-pin DIL	UV EPROM	+5 V at 50 mA
27C64	64K	64K words × 8 bit	28-pin DIL	CMOS UV EPROM	+5 V at 30 mA
613256	256K	32K words × 8 bit	28-pin DIL	Mask ROM	+5 V at 50 mA
AT28C256	256K	32K words × 8 bit	28-pin DIL	CMOS EEPROM	+5 V at 50 mA
AM29F010	1M	128K words × 8 bit	32-pin PLCC	Flash EPROM	+5 V at 30 mA
AT27010	1M	128K words × 8 bit	32-pin PLCC	OTP EPROM	+5 V at 25 mA

6.4 Programs and software

A program is simply a sequence of instructions that tells the computer to perform a particular operation. As far as the microprocessor is concerned, each instruction comprises a unique pattern of binary digits. It should, however, be noted that the computer may also require data and/or addresses of memory locations in order to fulfil a particular function. This information must also be presented to the microprocessor in the form of binary bit patterns. Clearly it is necessary for the processor to distinguish between the instruction itself and any data or memory address which may accompany it. Furthermore, instructions must be carried out in strict sequence. It is not possible for the microprocessor to execute more than one instruction at a time!

A computer program may exist in one of several forms although ultimately the only form usable by the processor is that which is presented in binary. This is unfortunate since most humans are not very adept in working in binary. The following program fragment presented in binary is for an Intel x86 family or Pentium processor:

```
10111000
00000001
00000000
10111011
00000010
00000001
11011000
10001001
11000001
11110100
```

It is hard to tell that the program takes two numbers, adds them together, and stores the result in one of the CPU registers. Clearly a more meaningful program representation is required!

An obvious improvement in making a program more understandable (at least for humans) is to write the program in a number base with which we are familiar. For this purpose, we could choose octal (base 8), decimal (base 10), or hexadecimal (base 16). The decimal equivalent of the above program code would, for example, be:

184, 1, 0, 187, 2, 0, 1, 216, 137, 193, 244

There is clearly not much improvement, although it is somewhat easier to recognize errors. We would, of course, still require some means of converting the decimal version of the program into binary so that the program makes sense to the processor. A better method, and one which we introduced you to in Chapter 2, uses hexadecimal. Hopefully you will recall that we can easily convert eight bits of binary code into its equivalent hexadecimal by arranging the binary values into groups of four bits and then converting each four-bit group at a time. For example, taking the first byte of the program that we met earlier:

$1011 = B$ and $1000 = 8$

therefore $10111000 = B8$ when written in hex.

Similarly:

$0000 = 0$ and $0001 = 1$

therefore $00000001 = 01$ when written in hex.

The simple addition program that we met earlier in its binary form, takes on the following appearance when written in hex:

B8, 01, 00, BB, 02, 00, 01, D8, 89, C1, F4

Whilst it still does not make much sense (unless you just happen to be familiar with x86 machine language programming!) a pattern can be easily recognized with the program written in this form. Furthermore, the conversion to and from binary has been a relatively simple matter.

6.4.1 Instruction sets

An instruction set is the name given to the complete range of instructions that can be used with any particular microprocessor. It should be noted that, although there are many similarities, instruction sets are unique to the particular microprocessor concerned. In some cases manufacturers have attempted to ensure that their own product range of microprocessors share a common sub-set of instructions. This permits the use of common software, simplifies development, and helps make improvements in microprocessors more acceptable to equipment designers and manufacturers. The Intel Pentium processor, for example, was originally designed as an enhancement of the 8086, 80286, 80386, and 80486 processors. These processors share a common sub-set of instructions and thus a program written for the (now obsolete) 8086 processor will run on a Pentium-based system.

Sophisticated microprocessors offer a large number of instructions; so much so, in fact, that the sheer number and variety of instructions can often be bewildering. It is also rather too easy to confuse the power of a microprocessor with the number of instructions that are available from its instruction set; these two things are not always directly related.

Instructions are presented to the microprocessor in words that occupy one or more complete bytes (often one, two, three or four bytes depending upon the processor and whether it is an 8-bit or 16, 32 or 64-bit type). These instruction words are sent to a register within the processor that is known appropriately as the **Instruction Register**.

6.4.2 Mnemonics

Most people find instructions written in hex rather difficult to remember, even though one does get to know some of the more common hex codes after working with a particular processor over a period of time. What is needed, therefore, is a simple method of remembering instructions in a form that is meaningful to the programmer. To this end, microprocessor manufacturers provide us with mnemonics for their instruction codes. They are given in a shorthand form that has some minor variations from one manufacturer to the next. One of the most common instructions is associated with loading or moving data into a register. For an Intel processor, the instruction or **operation code** mnemonic for this operation is written MOV.

For the load instruction to be meaningful we need to tell the processor which register is to be loaded and with what. This information, which is known as an **operand**, must also be contained in the instruction. Thus, if we wished to load the 16-bit accumulator register (known as the **AX register**) of an Intel processor with a 16-bit data value equivalent to 01 hexadecimal, we would use the instruction:

```
MOV AX,01h
```

This simply means 'place a data value of 01 hex. in the AX register'. The 'h' simply indicates that we have specified the address in hexadecimal.

If, alternatively, we simply wished to copy the contents of the AX register into the CX register we would write:

```
MOV CX,AX
```

Notice how the destination register appears before the source register.

Each of these two instructions has its own hexadecimal representation, the first being:

B8, 01, 00

whilst the second is:

89, C1

Notice how the first instruction consists of three hex bytes whilst the second comprises only two bytes.

A program written in instruction code mnemonics is known as an **assembly language** program. The process of converting the mnemonics to their hexadecimal equivalents is known as assembly and, whilst it is possible to translate assembly language programs to hexadecimal by referring to instruction code tables (i.e. hand assembly) this is, to say the least, a somewhat tedious and repetitive task and one which is very much prone to error. Instead, we use a computer program to perform this task. This assembler program accepts **source code** written in assembly language and converts this to **object code** that the CPU can execute.

The simple addition program that we met on page 71 takes the following form when written in x86 assembly language:

```
MOV AX,0001  ; Load AX with 16-bit
             ; data
MOV BX,0002  ; load BX with 16-bit
             ; data
ADD  AX,BX   ; add the two values
             ; together
MOV  CX,AX   ; place the result in CX
HLT          ; and stop
```

Note that we have also added some comments (after the semi-colons) in order to clarify the operation of the source code. Comments can be useful for maintenance or subsequent debugging.

6.4.3 Instructions

The available instructions within a specific computer instruction set (see page 71) can be grouped into a number of categories according to their function. Table 6.5 shows a subset of the x86 instruction set organised by their class or group. Most of the basic instructions can be used in several different ways according to the way in which they reference the source and/or destination data. Note also that the individual registers (see Chapter 7) may be either general purpose or may have some specific purpose (such as acting as an accumulator or as a counter).

The following x86 data movement instructions give some idea of the range of options available:

MOV AL,data	Moves 8-bit immediate data into the least-significant byte of the accumulator, AL.
MOV AH,data	Moves 8-bit immediate data into the most-significant byte of the accumulator, AH.
MOV AX,reg	Copies the contents of the specified 16-bit register to the 16-bit extended accumulator (AX).
MOV AH,reg	Copies the byte present in the specified 8-bit register to the 8-bit register, AH.
MOV AX,[addr]	Copies the 16-bit word at the specified address into the general-purpose base register, AX.

Finally, as a further example of the use of assembly language mnemonics, the following short code fragment reads an input port and transfers 16K bytes of data from the port to a series of memory locations, starting at address 7000 hex.:

```
MOV DX,0300h  ; port used for input
MOV AX,7000h  ; address to start data
MOV DS,AX     ; set up data segment
MOV DI,0      ; first location
MOV CX,04FFFh ; set data size to 16K
IN AL,DX      ; get a byte of data
MOV [DI],AL   ; and save it
INC DI        ; move to next location
CALL delay    ; wait a short time
LOOP getdata  ; and go around again
RET           ; return to main prog.
```

Table 6.5 A subset of Intel processor instructions

Data transfer instructions	
MOV	Move byte or word to register or memory
IN	Input byte or word from a specified port
OUT	Output byte or word to a specified port
PUSH	Push word onto stack
POP	Pop word off stack
XCHG	Exchange byte or word
XLAT	Translate byte using a look-up table
Logical instructions	
NOT	Logical NOT of byte or word
AND	Logical AND of byte or word
OR	Logical OR of byte or word
XOR	Logical exclusive-OR of byte or word
Shift and rotate instructions	
SHL	Logical shift left of byte or word
SHR	Logical shift right of byte or word
ROL	Rotate left of byte or word
ROR	Rotate right of byte or word
Arithmetic instructions	
ADD	Add byte or word
SUB	Subtract byte or word
INC	Increment byte or word
DEC	Decrement byte or word
NEG	Negate byte or word (two's complement)
CMP	Compare byte or word
MUL	Multiply byte or word (unsigned)
DIV	Divide byte or word (unsigned)
Transfer instructions	
JMP	Unconditional jump
JE	Jump if equal
JZ	Jump if zero
LOOP	Loop unconditional
LOOPE	Loop if equal
LOOPZ	Loop if zero
Subroutine and interrupt instructions	
CALL, RET	Call, return from procedure
Control instructions	
PUSHF	Push flags onto stack
POPF	Pop flags off stack
ESC	Escape to external processor interface
LOCK	Lock bus during next instruction
NOP	No operation (do nothing)
WAIT	Wait for signal on TEST input
HLT	Halt processor

6.5 Backplane bus systems

The computer systems used on modern aircraft are becoming increasingly sophisticated, frequently requiring powerful 32 and 64-bit processors capable of manipulating large amounts of data in a small time. In addition, many avionic systems may need several processors operating concurrently, large memories, and complex I/O sub-systems.

In order to meet these needs, today's avionic computer systems use a **backplane bus** which generally comprises of a number of identical sized cards mounted in a frame and linked together at the rear of the card frame by tracks on a printed circuit board mounted at right angles to the cards. This form of bus provides a means of linking together the sub-system components of a larger computer system. Each sub-system component consists of a PCB card and a number of VLSI devices. Such a system may incorporate several microprocessors (each of which with its own support devices and local bus system) or may involve just a single microprocessor (again with support devices and its own local bus) operating in conjunction with a number of less intelligent supporting cards.

Backplane bus systems are inherently flexible and the modular nature of the bus provides a means of easily changing and upgrading a system to meet new and changing demands. A high degree of modularity also allows cards to be exchanged when faults develop or interchanged between systems for testing. The user is thus able to minimize down time and need not be concerned with board-level servicing as cards can be returned to manufacturers or their appointed (and suitably equipped) service agents.

At the conceptual level, the functional elements of a bus system can be divided into **bus masters** (intelligent controlling devices that can generate bus commands), **bus slaves** (devices that generally exhibit less intelligence and cannot themselves generate bus commands and exercise control over the bus), and **intelligent slaves** (these are slaves that have their own intelligent controlling device but which do not themselves have the ability to place commands on the bus).

The ability to support more than one bus master is clearly a highly desirable facility and

one which allows us to exploit the full potential of a bus system. Systems that can do this are referred to as **multiprocessing systems**. Note that, whilst several bus masters can be connected to a bus, only one of them can command the bus at any time. The resources offered by the slaves can be shared between the bus masters.

In systems that support more than one potential bus master, a system of **bus arbitration** must be employed to eliminate possible contentions for control of the bus. Several techniques are used to establish bus priority and these generally fall into two classes, serial and parallel.

In a serially arbitrated system, bus access is granted according to a priority that is based on the physical slot location. Each master present on the bus notifies the next lower priority master when it needs to gain access to the bus. It also monitors the bus request status of the next higher priority master. The masters thus pass bus requests on from one to the next in a daisy-chain fashion.

In a parallel arbitrated system, external hardware is used to determine the priority of each bus master. Both systems have their advantages and disadvantages and some bus standards permit the use of both techniques.

6.5.1 The VME bus

One of the popular local computer bus systems used in aircraft today is the VersaModule Eurocard (VME) bus. The standard resulted from a joint effort between semiconductor manufacturers Mostek, Motorola and Signetics to establish a framework for 8, 16, and 32-bit parallel-bus computer architectures capable of supporting both single and multiple processors. The bus provides fast data transfer rates and uses two DIN 41612 indirect connectors fitted to each **bus c**ard in order to provide access to the full 32-bit bus. The basic VME standard defines four main buses present within a system:

- **data transfer bus** which provides a means of transferring data between bus cards
- **priority interrupt bus** which provides a means of alerting processors to external events in real-time and prioritising them so the most important is dealt with first

- **arbitration bus** which provides a means of determining which processor or bus master has control of the bus at any particular time
- **utility bus** which provides a means of distributing power and also synchronizing power-up and power-down operations.

Various sub-architectures, including the use of a local **mezzanine bus,** are possible within the overall VME framework. Further standards apply to these systems.

Figure 6.9 A VME bus processor card (note the two indirect edge connectors along the lower edge of the card)

Test your understanding 6.3

1. State THREE advantages of backplane bus systems.

2. Explain what is meant by the following terms:
   (a) backplane bus
   (b) bus master
   (c) bus slave.

3. Explain the need for bus arbitration in a multiprocessor system.

4. Describe two methods of bus arbitration.

5. Give and example of a commonly used backplane bus standard.

6.6 Some examples of aircraft computer systems

To help put this chapter into context, we shall conclude with a brief overview of two simple computer systems found in large transport aircraft. The information provided here is designed to show the different approaches to computer architecture and the interconnections made with other aircraft avionic systems.

6.6.1 Clock computer

The Captain and First Officer's clocks are required to provide various functions, not just indicating the current time. This makes them ideal candidates for the use of a microprocessor in a simple, yet essential computer system.

Figure 6.10 shows the simplified block schematic of the clock computer. The unit is powered from the aircraft's 28 V DC hot-battery bus and the functions that is provides are:

- Continuous display of Greenwich Mean Time (GMT)
- Secondary display of elapsed time or chronograph time (as selected by the Captain or First Officer)
- GMT output to the Flight Management Computer (FMC) via the ARINC 429 bus.

Inputs to the clock system consist of a number of switches that can be used for setting the clock parameters, for changing the clock mode, and freezing the clock display. The master timing signal for the clock is derived from an accurate crystal oscillator.

6.6.2 Aircraft integrated data system

The aircraft integrated data system (AIDS) collects and records operational data to permit detailed analysis of aircraft performance and maintenance requirements. The data is acquired by the digital flight data acquisition unit (DFDAU) and preserved for future reference by a quick access recorder (QAR). This latter device comprises a cartridge drive that records the data

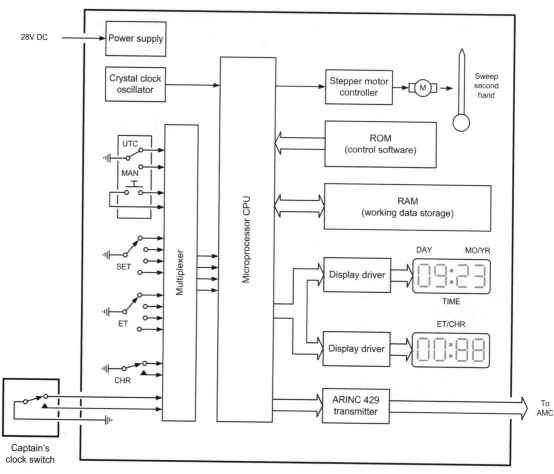

Figure 6.10   Clock computer

parameters on magnetic tape. The tape cartridge has a capacity of 14 hours of data storage. The QAR is able to search for and replay data from a specified GMT reference before subsequently returning to the current recording position.

A data management entry panel (DMEP) provides keyboard input of hexadecimal data and a limited set of commands (such as ENTER, CLEAR, PRINT, etc). Data and status information is displayed on an LCD display located on the Control Display Unit (CDU). The data recording system within the AIDS system is shown in Fig. 6.9. Interconnection and data exchange with other aircraft systems is via the ARINC 629 and ARINC 717 buses.

Key Point

Modern transport aircraft make extensive use of distributed computer systems. Some of these systems are based on the use of a single microprocessor device together with its support devices connected in a local bus arrangement. More complex systems use backplane bus systems in which multiple processors have shared access to the system's resources. Interconnection and data exchange between the various avionic computer systems is based on the use of serial bus standards such as ARINC 429, ARINC 629 and ARINC 717.

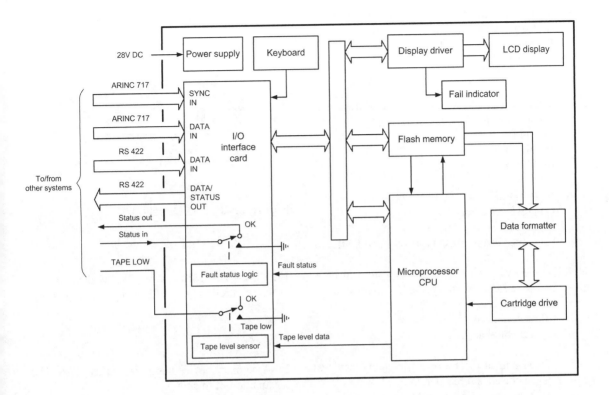

Figure 6.11 AIDS data recorder sub-system

6.7 Multiple choice questions

1. The feature marked Q in Figure 6.12 is:
   (a) RAM
   (b) ROM
   (c) I/O.

2. In Figure 6.12, the address is the feature marked:
   (a) T
   (b) U
   (c) V.

3. In Figure 6.12, which feature is responsible for providing read/write memory?
   (a) Q
   (b) R
   (c) S.

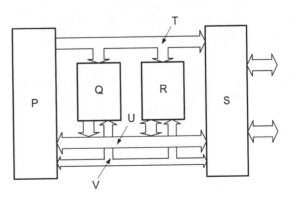

Figure 6.12 See Questions 1, 2 and 3

4. Which computer bus is used to specify memory locations?
   (a) address bus
   (b) control bus
   (c) data bus.

5. What is the largest hexadecimal address that can appear on a 24-bit address bus?
   (a) FFFF
   (b) FFFFF
   (c) FFFFFF.

6. Bus arbitration is required in order to:
   (a) prevent memory loss
   (b) avoid bus contention
   (c) reduce errors caused by noise and EMI.

7. A memory device in which any item of data can be retrieved with equal ease is known as:
   (a) parallel access
   (b) random access
   (c) sequential access.

8. Which one of the following memory devices can be erased and reprogrammed?
   (a) Mask programmed ROM
   (b) OTP EPROM
   (c) EPROM.

9. The processor instruction HLT is classed as:
   (a) a control instruction
   (b) a data transfer instruction
   (c) a logical instruction.

10. The VME bus standard uses:
    (a) a single 32-way DIN 41612 connector
    (b) two 32-way DIN 41612 connectors
    (c) a 100-way PCB edge connector.

11. What type of memory uses the principle of charge storage?
    (a) bipolar static RAM
    (b) MOS static memory
    (c) MOS dynamic memory.

12. The memory cell shown in Figure 6.13 is:
    (a) an MOS static cell
    (b) a MOS dynamic cell
    (c) a bipolar static cell.

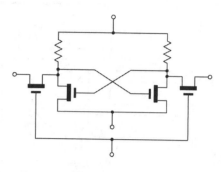

Figure 6.13   See Question 12.

13. How many 16K × 4 bit DRAM devices will be required to provide 32 K bytes of storage?
    (a) 2
    (b) 4
    (c) 8.

14. A memory device has a pin marked CAS. The function of this pin is:
    (a) chip active select
    (b) control address signal
    (c) column address select.

15. A semiconductor memory consists of 256 rows and 256 columns. The capacity of this memory will be:
    (a) 256 bits
    (b) 512 bits
    (c) 64K bits.

16. In the assembly language instruction MOV AX,07FEh the operation code is:
    (a) MOV
    (b) AX
    (c) 07FE.

17. A bus arbitration system based on the physical location of cards is referred to as:
    (a) serial arbitration
    (b) parallel arbitration
    (c) sequential arbitration.

<table>
<tr><td>**Chapter 7**</td><td># The CPU</td></tr>
</table>

The **microprocessor** central processing unit (CPU) forms the heart of any microprocessor or microcomputer system computer and, consequently, its operation is crucial to the entire system. The primary function of the microprocessor is that of fetching, decoding, and executing instructions resident in memory. As such, it must be able to transfer data from external memory into its own internal registers and vice versa. Furthermore, it must operate predictably, distinguishing, for example, between an operation contained within an instruction and any accompanying addresses of read/write memory locations. In addition, various system housekeeping tasks need to be performed including being able to suspend normal processing in order to responding to an external device that needs attention.

7.1 Internal architecture

The main parts of a microprocessor CPU are:

(a) **registers** for temporary storage of addresses and data;
(b) an **arithmetic logic unit** (ALU) that is capable of performing arithmetic and logical operations;
(c) a unit that receives and decodes instructions;
(d) a means of controlling and timing operations within the system.

Figure 7.1 shows the principal internal features of a typical 8-bit microprocessor as well as the data paths that link them together. Because this diagram is a little complex, we will briefly explain each of these features and what they do.

7.1.1 Accumulator

The accumulator functions both as a source and as a destination register for many of the basic microprocessor operations. As a **source register**

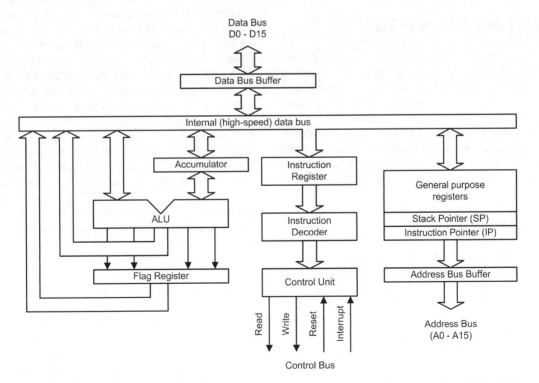

Figure 7.1   Internal architecture of a basic 8-bit microprocessor CPU

it contains the data that will be used in a particular operation whilst as a **destination register** it will be used to hold the result of a particular operation. The accumulator (or **A register**) features in a very large number of microprocessor operations, consequently more reference is made to this register than any others.

7.1.2 Instruction register

The instruction register provides a temporary storage location in which the current microprocessor instruction is held whilst it is being decoded. Program instructions are passed into the microprocessor, one at time, through the data bus.

On the first part of each **machine cycle**, the instruction is fetched and decoded. The instruction is executed on the second (and subsequent) machine cycles. Each machine cycle takes a finite time (usually less than a microsecond) depending upon the frequency of the microprocessor's clock.

7.1.3 Data bus (D0 to D7)

The external data bus provides a highway for data that links all of the system components (such as random access memory, read-only memory, and input/output devices) together. In an 8-bit system, the data bus has eight data lines, labelled D0 (the **least significant bit**) to D7 (**the most significant bit**) and data is moved around in groups of eight bits, or **bytes**. With a sixteen bit data bus the data lines are labelled D0 to D15, and so on.

7.1.4 Data bus buffer

The data bus buffer is a temporary register through which bytes of data pass on their way into, and out of, the microprocessor. The buffer is thus referred to as **bi-directional** with data passing out of the microprocessor on a **write operation** and into the processor during a **read operation**. The direction of data transfer is determined by the **control unit** as it responds to each individual program instruction.

7.1.5 Internal data bus

The internal data bus is a high-speed data highway that links all of the microprocessor's internal elements together. Data is constantly flowing backwards and forwards along the internal data bus lines.

7.1.6 General purpose registers

Many microprocessor operations (for example, adding two 8-bit numbers together) require the use of more than one register. There is also a requirement for temporarily storing the partial result of an operation whilst other operations take place. Both of these needs can be met by providing a number of general purpose registers. The use to which these registers are put is left mainly up to the programmer.

7.1.7 Stack pointer

When the time comes to suspend a particular task in order to briefly attend to something else, most microprocessors make use of a region of external random access memory (RAM) known as a **stack**. When the main program is interrupted, the microprocessor temporarily places in the stack the contents of its internal registers together with the address of the next instruction in the main program. When the interrupt has been attended to, the microprocessor recovers the data that has been stored temporarily in the stack together with the address of the next instruction within the main program. It is thus able to return to the main program exactly where it left off and with all the data preserved in its registers. The stack pointer is simply a register containing the address of the last used stack location.

7.1.8 Instruction pointer

As mentioned earlier in Chapter 6, computer programs consist of a sequence of instructions that are executed by the microprocessor. These instructions are stored in external random access memory (RAM) or read-only memory (ROM).

Instructions must be fetched and executed by the microprocessor in a strict sequence. By storing the address of the next instruction to be executed, the instruction pointer (or **program counter**) allows the microprocessor to keep track of where it is within the program. The program counter is automatically incremented when each instruction is executed.

7.1.9 Address bus buffer

The address bus buffer is a temporary register through which addresses (in this case comprising 16-bits) pass on their way out of the microprocessor. In a simple microprocessor, the address buffer is unidirectional with addresses placed on the address bus during both read and write operations. The address bus lines are labeled A0 to A15, where A0 is the least significant address bus line and A15 is the most significant address bus line. Note that a 16-bit address bus can be used to communicate with 65,536 individual memory locations. At each location a single byte of data is stored.

7.1.10 Control bus

The control bus is a collection of signal lines that are both used to control the transfer of data around the system and also to interact with external devices. The control signals used by microprocessors tend to differ with different types, however the following are commonly found:

READ   an output signal from the CPU that indicates that the current operation is a read operation

WRITE   an output signal from the CPU that indicates that the current operation is a write operation

RESET   a signal that resets the internal registers and initialises the instruction pointer program counter so that the program can be re-started from the beginning

IRQ   an interrupt request from an external device attempting to gain the attention of the CPU (the request may either be obeyed or ignored according to the

state of the microprocessor at the time that the interrupt request is received)

NMI   non-maskable interrupt (i.e. an interrupt signal that cannot be ignored by the microprocessor).

7.1.11 Address bus (A0 to A15)

The address bus provides a highway for addresses that links with all of the system components (such as random access memory, read-only memory, and input/output devices). In a system with a 16-bit address bus, there are sixteen address lines, labelled A0 (the least significant bit) to A15 (the most significant bit). In a system with a 32-bit address bus there are 32 address lines, labelled A0 to A31, and so on.

7.1.12 Instruction decoder

The instruction decoder is nothing more than an arrangement of logic gates that acts on the bits stored in the instruction register and determines which instruction is currently being referenced. The instruction decoder provides output signals for the microprocessor's control unit.

7.1.13 Control unit

The control unit is responsible for organising the orderly flow of data within the microprocessor as well as generating, and responding to, signals on the control bus. The control unit is also responsible for the timing of all data transfers. This process is synchronised using an internal or external clock signal (not shown in Fig. 7.1).

7.1.14 Arithmetic logic unit (ALU)

As its name suggests, the ALU performs arithmetic and logic operations. The ALU has two inputs (in this case these are both 8-bits wide). One of these inputs is derived from the Accumulator whilst the other is taken from the internal data bus via a temporary register (not shown in Fig. 7.1). The operations provided by the ALU usually include addition, subtraction, logical AND, logical OR, logical exclusive-OR, shift left, shift right, etc. The result of most ALU operations appears in the accumulator.

7.1.15 Status register

The result of an ALU operation is sometimes important in determining what subsequent action takes place. The status register (**flag register** or **condition code register**) contains a number of individual bits that are set or reset according to the outcome of an ALU operation. These bits are referred to as **flags**. The following flags are some typical examples of those provided by most microprocessors:

ZERO — the zero flag is set when the result of an ALU operation is zero

CARRY — the carry flag is set whenever the result of an ALU operation (such as addition) generates a carry bit (in other words, when the result cannot be contained within an 8-bit register)

INTERRUPT — the interrupt flag indicates whether external interrupts are currently enabled or disabled.

7.1.16 Clocks

The clock used in a computer system is simply an accurate and stable square wave generator. In most cases the frequency of the square wave generator is determined by a quartz crystal. A simple 4 MHz square wave clock oscillator (together with the clock waveform that is produces) is shown in Figure 7.2. Note that one complete clock cycle is sometimes referred to as a T-state.

Microprocessor central processing units sometimes have an internal clock circuit in which case the quartz crystal (or other resonant device) is connected directly to pins on the microprocessor chip. In Figure 7.3(a) an external clock is shown connected to a microprocessor whilst in Figure 7.3(b) an internal clock oscillator is used.

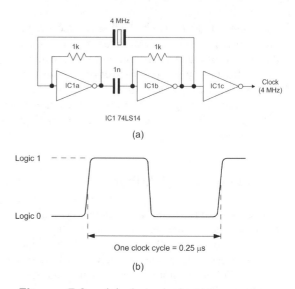

(a)

(b)

Figure 7.2 (a) A typical microprocessor clock circuit (b) waveform produced by the clock circuit

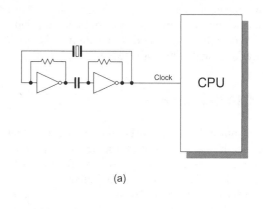

(a)

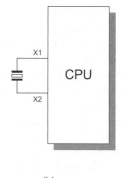

(b)

Figure 7.3 (a) An external CPU clock, and (b) an internal CPU clock

Test your understanding 7.1

Figure 7.6 shows the simplified internal architecture of an 8-bit microprocessor CPU. Use this diagram to complete the missing information in Table 7.1.

Test your understanding 7.2

A microprocessor CPU uses an 8 MHz clock. What would the time be for one T-state?

Test your understanding 7.3

In relation to the operation of a microprocessor CPU, briefly explain each of the following terms:

(a) interrupt request
(b) stack
(c) clock.

Test your understanding 7.4

Sketch a typical microprocessor clock waveform and indicate typical values for a 20 MHz clock.

Figure 7.4 The Z80 is a basic 8-bit microprocessor supplied in a 40-pin dual-in-line (DIL) package. The chip operates from a single +5 V supply.

7.2 Microprocessor operation

The majority of operations performed by a microprocessor involve the movement of data. Indeed, the program code (a set of instructions stored in ROM or RAM) must itself be fetched from memory prior to execution. The microprocessor thus performs a continuous sequence of instruction fetch and execute cycles. The act of fetching an instruction code (or operand or data value) from memory involves a read operation whilst the act of moving data from the microprocessor to a memory location involves a write operation, see Figure 7.5.

Each cycle of CPU operation is known as a machine cycle. Program instructions may require several machine cycles (typically between two and five). The first machine cycle in any cycle consists of an instruction fetch (the instruction code is read from the memory) and it is known as the M1 cycle. Subsequent cycles M2, M3, and so on, depend on the type of instruction that is being executed. This fetch-execute sequence is shown in Figure 7.7.

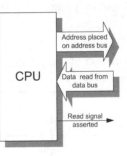

(a)

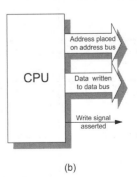

(b)

Figure 7.5 (a) Read, and (b) write operations

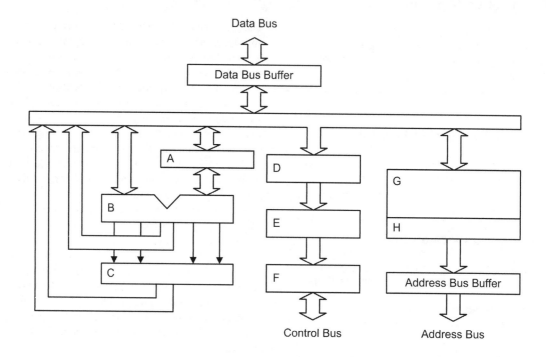

Figure 7.6 Internal architecture of an 8-bit processor (see Test your understanding 7.1)

Table 7.1 Function of the principal features in Figure 7.6 (see Test your understanding 7.1)

Name	Function	Feature (see Fig. 7.6)
Accumulator	Internal register that holds the result of an operation	A
	Device for performing arithmetic and logical operations	B
Control unit		F
Flag register	Internal register used to indicate the internal status of the processor	
General purpose registers	Internal registers used for the temporary storage of data during processing	
Instruction decoder	Logic circuit arrangement that decodes the current instruction code	
	Register that indicates the address of the next instruction	H
Instruction register	Internal register that stores the current instruction code	

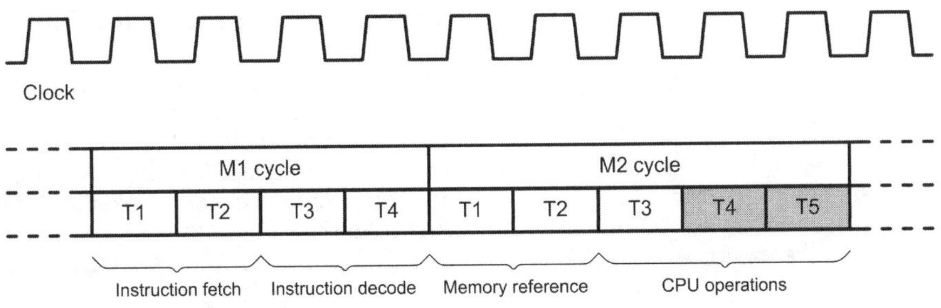

Figure 7.7 A typical timing diagram for a microprocessor CPU's fetch-execute cycle

Microprocessors determine the source of data (when it is being read) and the destination of data (when it is being written) by placing a unique address on the address bus. The address at which the data is to be placed (during a write operation) or from which it is to be fetched (during a read operation) can either constitute part of the memory of the system (in which case it may be within ROM or RAM) or it can be considered to be associated with input/output (I/O).

Since the data bus is connected to a number of VLSI devices, an essential requirement of such chips (e.g. ROM or RAM) is that their data outputs should be capable of being isolated from the bus whenever necessary. These chips are fitted with select or enable inputs that are driven by address decoding logic that ensures that external devices (ROM, RAM and I/O) never simultaneously attempt to place data on the bus.

The inputs of the address decoding logic are derived from one, or more, of the address bus lines. The address decoder effectively divides the available memory into blocks corresponding to a particular function (ROM, RAM, I/O, etc.). Hence, where the processor is reading and writing to RAM, for example, the address decoding logic will ensure that only the RAM is selected whilst the ROM and I/O remain isolated from the data bus.

Within the CPU, data is stored in several registers. Registers themselves can be thought of as a simple pigeon-hole arrangement that can store as many bits as there are holes available. Generally, these devices are can store groups of sixteen or thirty-two bits. Additionally, some registers may be configured as either one register

of sixteen bits or two registers of thirty-two bits.

Some microprocessor registers are accessible to the programmer whereas others are used by the microprocessor itself. Registers may be classified as either general purpose or dedicated. In the latter case a particular function is associated with the register, such as holding the result of an operation or signaling the result of a comparison.

A basic 8-bit microprocessor (the Z80) and its **register model** is shown in Fig. 7.8. Note that this microprocessor has six general purpose registers and that these are 8-bits in length. The registers can also be used 'end-on' so that, for example, the BC register pair can be used to hold 16-bit data. Note also, that the Z80's instruction pointer is referred to as the **program counter** and the status register is called the **flag register**. Note that different manufacturers use different names for these registers but their function remains the same.

7.2.1 ALU operation

The ALU can perform arithmetic operations (addition and subtraction) and logic (complementation, logical AND, logical OR, etc). The ALU operates on two inputs (eight, sixteen, thirty-two or sixty-four bits in length depending upon the CPU type) and it provides one output (again of eight, sixteen, thirty-two or sixty-four bits depending upon the CPU type).

The ALU status is preserved in the **flag register** so that, for example, an overflow, zero or negative result can be detected and the necessary action can then be taken to deal with this

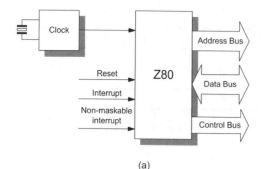

(a)

Main register set

Accumulator (A)	Flags (F)
(B)	(C)
(D)	(E)
(H)	(L)

Special purpose registers

Interrupt Vector (I)	Memory Refresh (R)
Index Register (IX)	
Index Register (IY)	
Stack Pointer (SP)	
Program Counter (PC)	

(b)

Figure 7.8 The Z80 CPU (showing some of its more important control signals and its register model)

eventuality. A typical example might be that of a program that needs to repeat an operation a set number of times until a zero result is obtained.

The control unit is responsible for the movement of data within the CPU and the management of control signals, both internal and external. The control unit asserts the requisite signals to read or write data as appropriate to the current instruction.

Test your understanding 7.5

With the aid of a timing diagram, explain the meaning of the term 'fetch-execute cycle'.

Test your understanding 7.6

The execution of a microprocessor instruction involves an M1 cycle comprising four T-states and an M2 cycle requiring five T-states. If the CPU clock is running at 20 MHz, how long will it take for the microprocessor to execute the complete instruction?

7.3 Intel x86 family

The original member of the x86 family was Intel's first true 16-bit processor which had 20 address lines that could directly address up to 1 MB of RAM. The chip was available in 5, 6, 8, and 10 MHz versions. The 8086 was designed with modular internal architecture. This approach to microprocessor design has allowed Intel to produce a similar microprocessor with identical internal architecture but employing an 8-bit external bus. This device, the 8088, shares the same 16-bit internal architecture as its 16-bit bus counterpart. Both devices were packaged in 40-pin DIL encapsulations. The CPU signal lines are described in Table 7.2.

The 8086/8088 can be divided internally into two functional blocks comprising an Execution Unit (EU) and a Bus Interface Unit (BIU), as shown in Figure 7.10. The EU is responsible for decoding and executing instructions, whilst the BIU pre-fetches instructions from memory and places them in an instruction queue where they await decoding and execution by the EU.

The EU comprises a general and special purpose register block, temporary registers, arithmetic logic unit (ALU), a flag (status) register, and control logic. It is important to note that the principal elements of the 8086 EU remain common to each of the subsequent members of the x86 family, but with additional registers with the more modern processors.

The BIU architecture varies according to the size of the external bus. The BIU comprises four segment registers and an instruction pointer, temporary storage for instructions held in the instruction queue, and bus control logic.

The **register model** and principal signals for the 8086 and x86 processors is shown in Fig. 7.9.

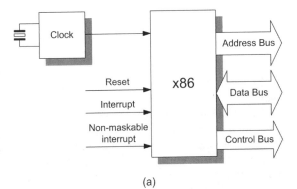

(a)

Main register set

AX	(AH)	(AL)
BX	(BH)	(BL)
CX	(CH)	(CL)
DX	(DH)	(DL)

Special purpose registers

Stack Pointer (SP)
Base Pointer (BP)
Destination Index (DI)
Source Index (SI)

Segment registers and instruction pointer

Code Segment (CS)
Data Segment (DS)
Stack Segment (SS)
Extra Segment (ES)
Instruction Pointer (IP)

(b)

Figure 7.9 The x86 CPU (showing some of its more important control signals and its register model

Note that the main registers can be used as either 8-bit or 16-bit registers. The 16-bit accumulator (register AX) for example, can be used as two 8-bit registers (AH and AL). In this case, Intel use the 'X' as an abbreviation for 'extended', 'H' for 'high' (i.e. the high byte), and 'L' for 'low' (i.e.

the low byte). Note also the use of **segment registers** for addressing. These registers allow the x86 processors to address memory outside the a 16-bit address space (without these registers the processor could only address a total of 64 Kbytes of memory and I/O). We will explain how this works in the next section.

7.3.1 Addressing

The 8086 has 20 address lines (see Table 7.2) and so provides for a physical 1 megabyte memory address range (memory address locations 00000 to FFFFF hex.). The I/O address range is 64 kilobytes (I/O addresses 0000 to FFFF hex.).

The actual 20-bit physical memory address is formed by shifting the segment address four zero bits to the left (adding four least significant bits), which effectively multiplies the Segment Register contents by 16. The contents of the Instruction Pointer (IP), Stack Pointer (SP) or other 16-bit memory reference is then added to the result. This process is illustrated in Figure 7 .11.

As an example of the process of forming a physical address reference, Table 7.3 shows the state of the 8086 registers after the RESET signal is applied. The instruction referenced (i.e. the first instruction to be executed after the RESET signal is applied) will be found by combining the Instruction Pointer (offset address) with the Code Segment register (paragraph address). The location of the instruction referenced is FFFF0 (i.e. F0000 + FFF0). Note that the PC's ROM physically occupies addresses F0000 to FFFFF and that, following a power-on or hardware reset, execution commences from address FFFF0 with a jump to the initial program loader.

7.3.2 80286, 80386 and 80486

The 80286 offers a total physical addressing of 16 megabytes but the chip also incorporates memory management capabilities that can map up to a gigabyte of virtual memory. Depending upon the application, the 80286 is up to six times faster than the standard 5 MHz 8086 while providing upward software compatibility with the 8086 and 8088 processors.

The 80286 had 15 16-bit registers, of which 14 are identical to those of the 8086. The additional Machine Status Word (MSW) register controls the operating mode of the processor and also records when a task switch takes place.

The bit functions within the MSW are summarized in Table 7.4. The MSW is initialized with a value of FFF0H upon reset, the remainder of the 80286 registers being initialized as shown in Table 7.3. The 80286 is packaged in a 68-pin JEDEC type-A plastic leadless chip carrier (PLCC).

The 80386 (or '386) was designed as a *full* 32-bit device capable of manipulating data 32 bits at a time and communicating with the outside world through a 32-bit address bus. The 80386 offers a 'virtual 8086' mode of operation in which memory can be divided into 1 megabyte chunks with a different program allocated to each partition.

The 80386 was available in two basic versions. The 80386SX operates internally as a 32-bit device but presents itself to the outside world through only 16 data lines. This has made the CPU extremely popular for use in low-cost systems which could still boast the processing power of a 80386 (despite the obvious limitation imposed by the reduced number of data lines, the

Table 7.2 8088/8086 signal lines

Signal	Function	Notes
AD0–AD7 (8088)	Address/data bus	Multiplexed 8-bit address/data lines
A8-A19 (8088)	Address bus	Non-multiplexed address lines
AD0–AD15 (8086)	Address/data bus	Multiplexed 16-bit address/data bus
A16-A19 (8086)	Address bus	Non-multiplexed address lines
S0-S7	Status lines	S0-S2 are only available in Maximum Mode and are connected to the 8288 Bus Controller. Note that status lines S3-S7 all share pins with other signals.
INTR	Interrupt line	Level-triggered, active high interrupt request input
NMI	Non-maskable interrupt line	Positive edge-triggered non-maskable interrupt input
RESET	Reset line	Active high reset input
READY	Ready line	Active high ready input
TEST	Test	Input used to provide synchronisation with external processors. When a WAIT instruction is encountered in the instruction stream, the CPU examines the state of the TEST line. If this line is found to be high, the processor waits in an 'idle' state until the signal goes low
QS0,QS1	Queue status lines	Outputs from the processor which may be used to keep track of the internal instruction queue
LOCK	Bus lock	Output from the processor which is taken low to indicate that the bus is not currently available to other potential bus masters
RQ/GT0-RQ/GT1	Request/Grant	Used for signalling bus requests and grants placed in the CL register.

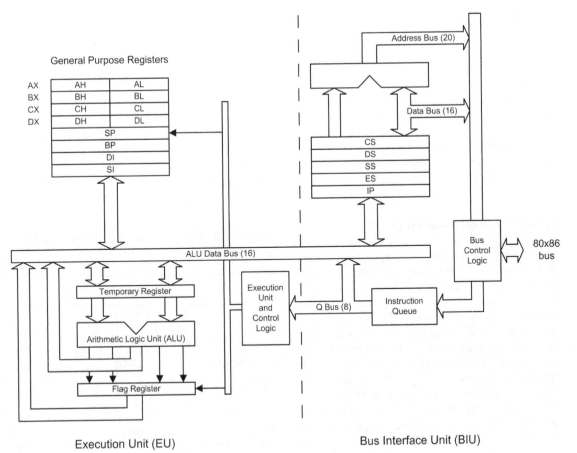

General Purpose Registers

Address Bus (20)

Data Bus (16)

Bus Control Logic

80x86 bus

ALU Data Bus (16)

Temporary Register

Execution Unit and Control Logic

Q Bus (8)

Instruction Queue

Arithmetic Logic Unit (ALU)

Flag Register

Execution Unit (EU)

Bus Interface Unit (BIU)

Figure 7.10 Internal architecture of the 8086

'SX' version of the 80386 runs at approximately 80% of the speed of its fully fledged counterpart).

The 80386 comprises a Bus Interface Unit (BIU), a Code Prefetch Unit, an Instruction Decode Unit, an Execution Unit (EU), a Segmentation Unit and a Paging Unit. The Code Prefetch Unit performs the program 'lookahead' function. When the BIU is not performing bus cycles in the execution of an instruction, the Code Prefetch Unit uses the BIU to fetch sequentially the instruction stream. The prefetched instructions are stored in a 16-byte 'code queue' where they await processing by the Instruction Decode Unit. The prefetch queue is fed to the Instruction Decode Unit which translates the instructions into microcode. These microcoded instructions are then stored in a

three-deep instruction queue on a first-in first-out (FIFO) basis. This queue of instructions awaits acceptance by the EU. Immediate data and opcode offsets are also taken from the prefetch queue.

The 80486 processor was not merely an upgraded 80386 processor; its redesigned architecture offers significantly faster processing speeds when running at the *same* clock speed as its predecessor. Enhancements include a built-in maths coprocessor, internal cache memory and cache memory control. The internal cache is responsible for a significant increase in processing speed. As a result, a '486 operating at 25 MHz can achieve a faster processing speed than a '386 operating at 33 MHz.

The '486 uses a large number of additional signals associated with parity checking (PCHK) and cache operation (AHOLD, FLUSH, etc.). The cache comprises a set of four 2-kilobyte blocks (128 x 16 bytes) of high-speed internal memory. Each 16-byte line of memory has a matching 21-bit 'tag'. This tag comprises a 17-bit linear address together with four protection bits. The cache control block contains 128 sets of seven bits. Three of the bits are used to implement the 'least recently used' (LRU) system for replacement and the remaining four bits are used to indicate valid data.

7.3.3 Interrupt handling

Interrupt service routines are subprograms stored away from the main body of code that are available for execution whenever the relevant interrupt occurs. However, since interrupts may occur at virtually any point in the execution of a main program, the response must be automatic; the processor must suspend its current task and save the return address so that the program can be resumed at the point at which it was left. Note that the programmer must assume responsibility for preserving the state of any registers which may have their contents altered during execution

of the interrupt service routine and restoring them after the interrupt has been serviced.

The Intel processor family uses a table of 256 4-byte pointers stored in the bottom 1 kilobyte of memory (addresses 0000H to 03FFH). Each of the locations in the Interrupt Pointer Table can be loaded with a pointer to a different interrupt service routine. Each pointer contains 2 bytes for loading into the Instruction Pointer (IP). This allows the programmer to place his or her interrupt service routines in any appropriate place within the 1 megabyte physical address space.

Table 7.3 Contents of the 8086 registers after a reset

Register	Contents (hex.)
Flag	0002
Instruction Pointer	FFF0
Code Segment	F000
Data Segment	0000
Extra Segment	0000
Stack Segment	0000

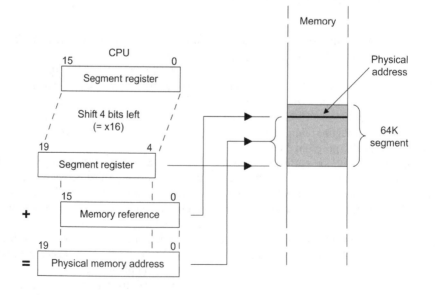

Figure 7.11 Process of forming a 20-bit address

Table 7.4 Bit functions of the 80286 machine status word

Bit	Name	Function
0	Protected mode (PM)	Enables protected mode and can only be cleared by asserting the RESET signal.
1	Monitor processor (MP)	Allows WAIT instructions to cause a 'processor extension not present' exception (Exception 7).
2	Emulate processor (EP)	Causes a 'processor extension not present' present' exception (Exception 7) on ESC instructions to allow emulation of a processor extension.
3	Task switched (TS)	Indicates that the next instruction using a processor extension will cause Exception 7 (allowing software to test whether the current processor extension context belongs to the current task).

7.4 The Intel Pentium family

Initially running at 60 MHz, the Pentium could achieve 100 million instructions per second. The original Pentium had (see Fig. 7.12) an architecture based on 3.2 million transistors and a 32-bit address bus like the 486 but a 64-bit external data bus. The chip was capable of operation at twice the speed of its predecessor, the '486. The Pentium was eventually to become available in 60, 66, 75, 90, 100, 120, 133, 150, 166, and 200 MHz versions. The first ones fitted Socket 4 boards whilst the rest fitted Socket 7 boards. The Pentium was **super-scalar** and could execute two instructions per clock cycle. With two separate 8K memory caches it was much faster than a '486 with the same clock speed.

The Pentium Pro incorporated a number of changes over the Pentium which made the chip run faster for the same clock speeds. Three instead of two instructions can be decoded in each clock cycle and instruction decoding and execution are decoupled, meaning that instructions can still be executed if one pipeline stops. Instructions could also be executed out of strict order. The Pentium Pro had an 8K level 1 cache for data and a separate cache for instructions. The chip was available with up to 1 MB of onboard L2 cache which further increased data throughput. The architecture of the Pentium Pro was optimized for 32-bit code but the chip would only run 16-bit code at the same speed as its predecessor.

Optimized for 32-bit applications, the Pentium II had 32 KB of L1 cache (16 KB each for data and instructions) and had a 512 KB of L2 cache on package. To discourage competitors from making direct replacement chips, this was the first Intel chip to make use of its patented 'Slot 1'. The Intel Celeron was a cut down version of Pentium II aimed primarily at the laptop computer market. The chip was slower as the L2 cache had been removed. Later versions were supplied with 128 KB of L2 cache.

The Pentium III was released in February 1999 and first made available in a 450 MHz version supporting 100 MHz bus. The latest Pentium IV architecture is based on its new 'NetBurst' architecture that combines four technologies; Hyper Pipelined Technology, Rapid Execution Engine, Execution Trace Cache and a 400MHz system bus.

The Pentium IV processor is available at speeds ranging from 1.70 GHz to 2.80 GHz with system bus speeds of 400 MHz and 533 MHz (the latter delivering a staggering 4.2 GB of data-per-second into and out of the processor). This performance is accomplished through a physical signaling scheme of quad pumping the data transfers over a 133 MHz clocked system bus and a buffering scheme allowing for sustained 533 MHz data transfers.

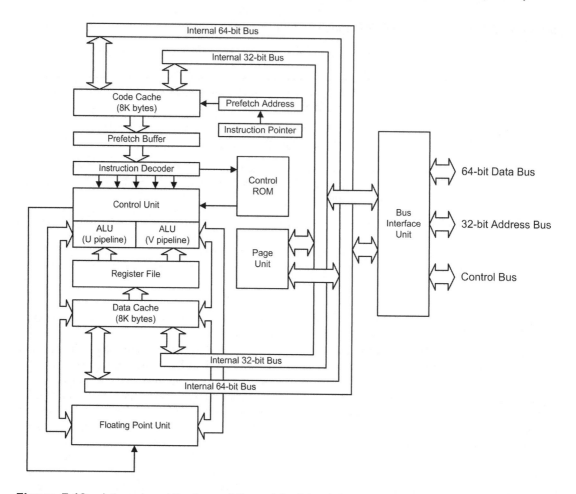

Figure 7.12 Internal architecture of the original Intel Pentium processor. Note the use of a separate bus interface unit (BIU)

7.5 AMD 29050

The AMD 29050 (and other 29K family members) have been extremely popular with avionics manufacturers over the past decade and many of these processors have been employed in critical applications. Despite its popularity in the aerospace sector, AMD discontinued the 29K series completely in 1995 due mainly to the development of the AMD K5 and K6 processors for mass-market applications. Further development of the high-end 29050 processor passed to Honeywell who used the 29050 core as the basis of an application specific integrated circuit (ASIC) which is currently used in its

Versatile Integrated Avionics (VIA) package. The VIA provides a variety of integrated avionics functions, including data logging, data display, and control of critical subsystems.

Each section of the VIA takes the form of a Core Processing Module (CPM) and each CPM is based on a pair of processors (either AMD 29050 or HI-29KII in the later Honeywell ASIC version). The CPUs are set-up in a system known as voting. All outputs (address, data and control) are passed to a comparator, if each CPU sends out the same signals then the system continues. If the signals do not match then a fault condition is present and the system enters an error mode in which it discontinues processing, logs the error and then attempts to recover. The 29050 (and its Honeywell derivative) is currently found on the Boeing 717, 737-600/700/800, 777 and the Federal Express MD-80.

The instruction set of the 29K family was designed to closely match the internal representation of operations generated by optimizing compilers. Instruction execution times are not burdened by redundant instruction formats and options found on other microprocessors.

The 29K family have a number of useful features that add to their attractiveness for powerful **embedded applications** such as those found in an aircraft. The 29K family is based on Reduced Instruction Set Computer (RISC) architecture. Essentially this means that relatively few instructions are used and that these instructions are kept extremely simple. These small, simple instructions are usually executed very much faster than longer, more complex ones. Furthermore, since most computer programs involve executing simple instructions over and over again, by making these instructions very efficient it is possible to make significant improvements to the overall speed of execution. Note that RISC is not just to do with the number of instructions that are available but rather it is to do with ensuring that the instructions that are used are very straightforward and execute in the quickest possible time.

Another important feature of the 29K family is the use of **instruction pipelining**. Each 29K processor has a four-stage pipeline consisting of first a fetch stage, followed by an instruction decode, then the execute and write-back stages. Instructions (with a few exceptions) execute in a single-cycle. Processors in the 29K family also make use of other techniques usually associated with RISC, such as delayed branching to keep the processor fed with a continuous stream of instructions and to prevent the pipeline stalling.

The high-end members of the 29K family (including the 29050 used in critical avionic applications) use a three bus Harvard architecture (see Fig. 7.14) rather than the simpler two bus (plus control) structure used with simpler processors. The key feature of the three bus architecture is the separation of instructions from the address bus. Instructions are fed to the processor by means of a separate **instruction bus**

Key Point

Modern integrated avionic systems demand processors that are both powerful and highly reliable. The core processor modules (CPM) of the Boeing 777 integrated avionics system (see Fig. 7.13) use pairs of processors operating in a self-checking 'Lockstep' configuration. The Boeing 777 incorporates Honeywell's Airplane Information Management System (AIMS) which is based on CPM fitted with pairs of AMD 29050 (AIMS-1) or HI-29KII (AIMS-2) processors. These systems work by continuously comparing all processor inputs and outputs (address, data, instructions). If a difference is detected, an error is flagged, all processing is halted (so that erroneous data is not propagated to other avionic systems) and the error is logged before a recovery is attempted.

---

from either the instruction read/write memory or the instruction ROM. An additional **bus bridge** is usually incorporated in order to provide a means of supplying instruction information from the data bus (as shown in Fig. 7.14).

Test your understanding 7.7

Briefly explain the meaning of the following terms:
(a) segmented addressing
(b) instruction pipeline.

Test your understanding 7.8

Explain how the Boeing 777 AIMS uses two identical microprocessors to implement a critical avionics system.

Test your understanding 7.9

With the aid of a diagram, explain how a three bus architecture can be used to improve processor performance.

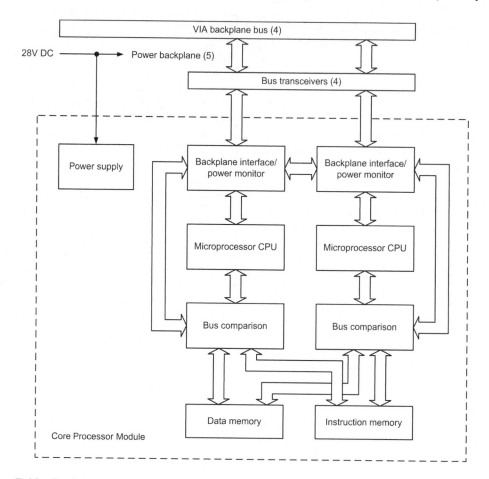

Figure 7.13   Basic core processor module used in the Boeing 777 Honeywell AIMS

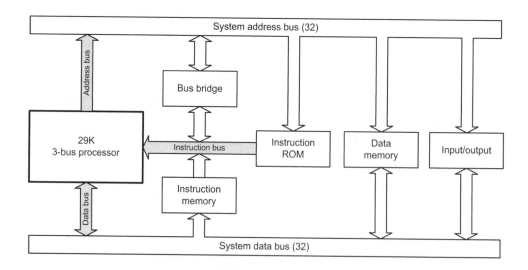

Figure 7.14   Three bus architecture of a high-end 29K system

7.6 Multiple choice questions

1. In the diagram of a CPU shown in Fig. 7.15, the accumulator is the feature marked:
   (a) A
   (b) B
   (c) C.

2. In the diagram of a CPU shown in Fig. 7.15, which feature indicates the current status of the processor?
   (a) A
   (b) B
   (c) C.

3. In the diagram of a CPU shown in Fig. 7.15, which feature performs logical operations?
   (a) A
   (b) B
   (c) C.

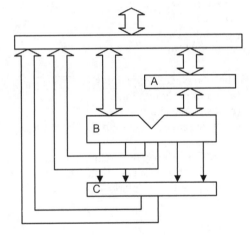

Figure 7.15 See Questions 1, 2 and 3

4. The stack is used for:
   (a) permanent storage of data
   (b) temporary storage of data
   (c) temporary storage of programs.

5. The stack is a structure located:
   (a) in the ALU
   (b) in the general purpose registers
   (c) in external read/write memory.

6. A byte of data is to be inverted. This task is performed by:
   (a) the ALU
   (b) the instruction register
   (c) the instruction decoder.

7. Data can be transferred from one bus to another by means of:
   (a) a bus cycle
   (b) a bus buffer
   (c) a bus bridge.

8. The output of the instruction pointer appears on:
   (a) the address bus
   (b) the data bus
   (c) the control bus.

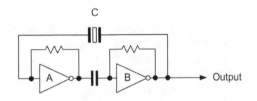

Figure 7.16 See Question 9, 10 and 11

9. The output of the circuit shown in Fig. 7.16 will be:
   (a) a square wave
   (b) a sine wave
   (c) a series of narrow pulses.

10. Which component in Fig. 7.16 determines the frequency of the output:
    (a) A
    (b) B
    (c) C.

11. A typical application for the circuit shown in Fig. 7.16 is:
    (a) a bus interface
    (b) a clock generator
    (c) a read/write memory.

12. The CPU data bus buffer is:
    (a) unidirectional
    (b) bidirectional
    (c) omnidirectional.

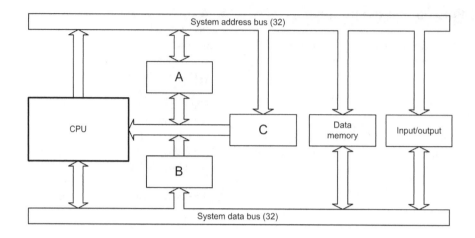

Figure 7.17 See Questions 20 and 21

13. Another way to describe the CPU register that acts as an instruction pointer is:
    (a) the program counter
    (b) the stack pointer
    (c) the instruction register.

14. In which cycle is an instruction fetched and decoded:
    (a) M0
    (b) M1
    (c) M2.

15. If a microprocessor clock runs at 50 MHz, an instruction requiring 11 T-states will execute in a time of:
    (a) 220 ns
    (b) 440 ns
    (c) 2.2 ms.

16. An external device can gain the attention of the CPU by generating a signal on the:
    (a) R/W line
    (b) RESET line
    (c) IRQ line.

17. When executing the assembly language instruction MOV AX,07FEh the operation code will be transferred to:
    (a) the accumulator
    (b) the instruction register
    (c) the instruction pointer.

18. After executing the assembly language instruction MOV AX,07FEh the binary data in the AL register will be:
    (a) 11101111
    (b) 00000111
    (c) 11111110.

19. The advantage of pipelining is:
    (a) easier programming
    (b) faster execution times
    (c) more reliable operation.

20. Figure 7.17 shows the architecture of a three bus microprocessor system. The instruction ROM is the feature marked:
    (a) A
    (b) B
    (c) C.

21. Figure 7.17 shows the architecture of a three bus microprocessor system. The bus bridge is the feature marked:
    (a) A
    (b) B
    (c) C.

22. Segmentation is used with x86 processors in order to:
    (a) increase the speed of processing
    (b) set up an instruction pipeline
    (c) extend the addressing range.

Chapter 8 Integrated circuits

Considerable cost savings can be made by manufacturing all of the components required for a particular circuit function on one small slice of semiconductor material (usually silicon). The resulting **integrated circuit** may contain as few as 10 or more than 100,000 active devices (transistors and diodes). With the exception of a few specialised applications (such as amplification at high power levels) integrated circuits have largely rendered conventional circuits (i.e. those based on discrete components such as individually packaged resistors, diodes and transistors) obsolete.

Integrated circuits can be divided into two general classes, linear (analogue) and digital. Typical examples of linear integrated circuits are operational amplifiers whereas typical examples of digital integrated circuits are the logic gates and microprocessors that you met in the earlier chapters.

It's worth noting that a number of integrated circuit devices bridge the gap between the analogue and digital world. Such devices include analogue to digital converters (ADC), digital to analogue converters (DAC), and timers. Table 8.2 outlines the main types of integrated circuit.

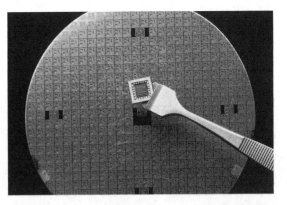

Figure 8.1 Multiple integrated circuit chips formed on a silicon wafer

8.1 Scale of integration

The relative size of a digital integrated circuit (in terms of the number of logic gates or equivalent devices that it contains) is often referred to as its **scale of integration**. The terminology shown in Table 8.1 is commonly used to describe the scale of these circuits.

Key Point

VLSI integrated circuits contain many thousands of components fabricated on a small piece of silicon. Each of these 'chips' can replace very large numbers of conventional components. Typical examples of VLSI devices are microprocessors and memory devices.

Table 8.1   Scale of integration

Scale of integration	Abbreviation	Number of logic gates *	Typical examples
Small	SSI	1 to 10	Basic logic (AND, OR, NAND, NOR, etc)
Medium	MSI	10 to 100	Bus buffers and transceivers; encoders and decoders, small programmed logic arrays
Large	LSI	100 to 1,000	Large gate arrays; small memory devices
Very large	VLSI	1,000 to 10,000	Large memory devices; small microprocessors
Ultra large	ULSI	More than 100,000	Large microprocessors

* or circuits of equivalent complexity

Table 8.2 Types of integrated circuits

Digital	
Logic gates	Digital integrated circuits that provide logic functions such as AND, OR, NAND and NOR
Microprocessors	Digital integrated circuits that are capable of executing a sequence of programmed instructions. Microprocessors are able to store digital data whilst it is being processed and to carry out a variety of operations on the data, including comparison, addition and subtraction
Memory devices	Integrated circuits (e.g. RAM and EPROM) used to store digital information
Analogue	
Operational amplifiers	Integrated circuits designed primarily for linear operation and which form the fundamental building blocks of a wide variety of linear circuits such as amplifiers, filters and oscillators
Low-noise amplifiers	Linear integrated circuits that are designed so that they introduce very little noise which may otherwise degrade low-level signals
Voltage regulators	Linear integrated circuits that are designed to maintain a constant output voltage in circumstances when the input voltage or the load current changes over a wide range
Hybrid (combined digital and analogue)	
Timers	Integrated circuits that are designed primarily for generating signals that have an accurately defined time interval such as that which could be used to provide a delay or determine the time between pulses. Timers generally comprise several operational amplifiers together with one or more bistable devices
Analogue to digital converters (ADC)	Integrated circuits that are used to convert a signal in analogue form to one in digital form. A typical application would be where temperature is sensed using a thermistor to generate an analogue signal. This signal is then converted to an equivalent digital signal using an ADC and then sent to a microprocessor for processing
Digital to analogue converters (DAC)	Integrated circuits that are used to convert a signal in digital form to one in analogue form. A typical application would be where the speed of a DC motor is to be controlled from the output of a microprocessor. The digital signal from the microprocessor is converted to an analogue signal by means of a DAC. The output of the DAC is then further amplified before applying it to the field winding of a DC motor.

8.2 Fabrication technology

Integrated circuits are fabricated on wafers of very pure silicon. The layers of semiconductor material are fabricated by means of a photographic process using ultraviolet light and a series of masks corresponding to the doping required for each semiconductor layer. The resolution of the photographic process ultimately determines the number of devices that can be integrated into a single chip (currently the processes used by leading manufacturers allow for a manufacturing resolution of 90 nm, or less). Using this process, multiple integrated circuit devices are formed on a single wafer of up to 30 cm in diameter (see Fig. 8.1).

When the doping process is complete, the wafer is tested before being cut into small rectangular areas called dice. Non-functional die are marked and rejected whilst those that are functional move on to the next stage of the process which involves mounting the die into a package using aluminium (or gold) wires which are welded to pads formed around the edge of the die. These tiny wires form the connection from the die to the connecting or soldering pins on the package in which the die is finally mounted and sealed hermetically. Testing accounts for a significant proportion of the production cost and low yield on the more complex devices (VLSI and ULSI packages) can be problematic and, as a result, costs can be high.

Figure 8.2 Common packages used for integrated circuits (the two resistors have been included for size comparison)

8.3 Packaging and pin numbering

The earliest integrated circuits in the 1960s were packaged in ceramic flat packs. These were used in military and critical aerospace applications for several decades due to their small size and high reliability. Packaging of integrated circuits for domestic, consumer and industrial applications moved quickly to the popular dual in-line package (DIL or DIP). The first generation of DIL integrated circuits were supplied in ceramic packages but lower-cost plastic packages quickly became popular for non-critical applications, such as the first generation of personal computers.

With the advent of powerful 16 and 32-bit microprocessors in the 1980s, pin counts of VLSI circuits exceeded the practical limit for DIP packaging (around 68 pins) and so the pin grid array (PGA) and leadless chip carrier (LCC) or plastic leadless chip carrier (PLCC) packages were introduced.

Surface mount packaging first appeared in the early 1980s and became widespread a decade, or so, later. Surface mounted integrated circuits use connections that were more closely spaced with their connections formed as either gull-wing or J-lead, as exemplified by the small-outline integrated circuit (SOIC).

PGA packages are still in common use but, in the late 1990s, PQFP and TSOP packages were introduced as a more space-efficient solution for integrated circuits with a high pin count (several hundred pins, or more).

Figure 8.3 An 8-bit CMOS microprocessor supplied in an 80-pin plastic quad flat pack (PQFP)

Figure 8.4 An 81-pin integrated circuit mounted in a pin grid array (PGA) package

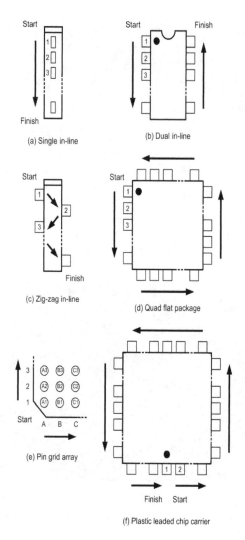

(a) Single in-line

(b) Dual in-line

(c) Zig-zag in-line

(d) Quad flat package

(e) Pin grid array

(f) Plastic leaded chip carrier

Figure 8.5 Pin connections and numbering for various integrated circuit packages (top view)

Table 8.3 summarises the various types of integrated circuit package whilst Figure 8.5 shows the corresponding pin numbering schemes. These are shown looking from the top of the device.

It is important to note that manufacturers often provide differently packaged versions of the same integrated circuit device. The different variants are usually distinguished by additional letters and/or numbers in the device coding. Figure 8.6 shows an example of two different styles of packaging used for a small MSI device.

Table 8.3 Integrated circuit packages and pin numbering

Abbrev.	Meaning	Figure 8.5
BGA	Ball grid array	(e)
DIL	Dual in-line	(b)
DIP	Dual in-line package	(b)
PGA	Pin grid array	(e)
PLCC	Plastic leadless chip carrier	(d, f)
PQFP	Plastic quad flat package	(d, f)
QFP	Quad flat package	(d, f)
SDIP	Shrink dual in-line package	(b)
SIP	Single in-line package	(a)
SIL	Single in-line	(a)
SO	Small outline	(b)
SOIC	Small outline integrated circuit	(b)
SOJ	Small outline J-lead	(b)
SOP	Small outline package	(b)
SSOP	Shrink small outline package	(b)
TSOP	Thin small outline package	(b)
ZIP	Zig-zag in-line package	(c)

Test your understanding 8.1

What do each of the following abbreviations stand for?

1. SSI
2. DIL
3. QFP
4. SOIC.

Test your understanding 8.2

Explain what is meant by a hybrid integrated circuit and give TWO examples of such devices.

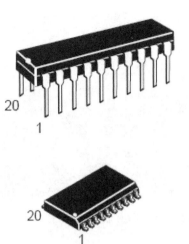

Figure 8.6 Pin numbering for 20-pin DIP and 20-pin SOIC versions of the same chip

8.4 Multiple choice questions

1. A standard logic gate manufactured in 1981 is likely to be supplied in a:
   (a) DIL package
   (b) PGA package
   (c) QFP package.

2. A standard logic gate is a typical example of:
   (a) an SSI device
   (b) an LSI device
   (c) a VLSI device.

3. Which one of the following scales of integration is the largest?
   (a) SSI
   (b) LSI
   (c) MSI.

4. A microprocessor is an example of::
   (a) SSI technology
   (b) MSI technology
   (c) VLSI technology.

5. When compared with DIP packaging, PLCC offers:
   (a) more pins
   (b) larger size
   (c) greater reliability.

6. Which one of the following integrated circuit packaging technologies requires that the chip should always be soldered in place?
   (a) DIL
   (b) SOIC
   (c) PGA.

7. Which of the following integrated circuit packaging technologies was the earliest to be used?
   (a) ceramic DIP
   (b) ceramic QFP
   (c) PLCC.

8. The connections from the pads on an integrated circuit die are:
   (a) soldered to the internal wire links
   (b) crimped to the internal wire links
   (c) welded to the internal wire links.

9. An integrated circuit bus transceiver contains 64 logic gates and buffers. This chip is an example of:
   (a) SSI
   (b) MSI
   (c) LSI.

10. The integrated circuit packages shown in Figure 8.7 are:
    (a) DIL
    (b) PGA
    (c) PLCC.

Figure 8.7 See Question 10

Figure 8.8   See Question 11

Figure 8.10   See Question 12

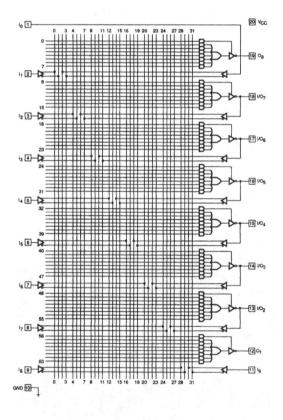

Figure 8.9   See Question 14

11. The integrated circuit package shown in
    Figure 8.8 is:
    (a) DIL
    (b) QFP
    (c) PLCC.

12. The number of the integrated circuit pin
    circled in Figure 8.10 is:
    (a) 1
    (b) 7
    (c) 8.

13. Production tests are performed on an
    integrated circuit wafer:
    (a) before cutting it into individual chips
    (b) after cutting it into individual chips
    (c) only when the chips are finally mounted.

14. The programmed logic device shown in
    Figure 8.9 is an example of:
    (a) SSI technology
    (b) MSI technology
    (c) LSI technology.

15. Each individual integrated circuit produced
    from a wafer is known as a:
    (a) blank
    (b) die
    (c) gate.

Chapter 9   MSI logic

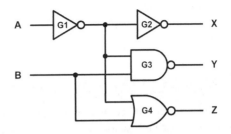

Figure 9.1 An example of determining the fan-in for a simple logic system. If all gates are standard logic devices then input A has a fan-in of 1 and input B has a fan in of 2

Medium scale integration (MSI) logic circuits are frequently used in avionic systems to satisfy the need for more complex logic functions such as those required for address decoding, code conversion, keyboard encoding/decoding, and the switching of logic signals between different bus connections. Some typical MSI applications include:

- address decoders
- priority encoders
- multiplexers and data selectors
- BCD to binary code converters
- BCD to seven-segment code converters

We shall begin this section by introducing two important concepts relating to the interconnection of large numbers of logic gates.

9.1 Fan-in and fan-out

When large numbers of logic gates are connected together the output of one logic gate may be connected to inputs on several other gates. In a practical logic circuit, the loading effect of these inputs needs to be taken into account.

Fan-out is defined as the maximum number of standard logic inputs (of the same logic family) that a logic gate can supply without risk of the logic levels falling out of specification.

Most standard TTL logic gates have a fan-out of 10. This means that they can drive up to 10 standard TTL inputs. Any more than this may result in unreliable operation because the gate is unable to source insufficient current when the output is in the 'high' state and sink insufficient current when the output is in the 'low' state. Either of these conditions is likely to compromise the logic levels required for successful operation.

Fan-in is a measure of the loading effect (expressed as the number of equivalent logic inputs of the same logic family) imposed by each of a gate's inputs. Thus, a standard logic gate

input has a fan-in of 1. Figure 9.1 shows an example of how the fan-in of a simple logic circuit is determined. Assuming that all of the logic gates in the circuit are standard devices (i.e. each of their inputs represents unity fan-in) then external input A has a fan-in of 1 (it only drives one input, G1) whilst external input B has a fan-in of 2 (it drives two standard inputs, one on G3 and one on G4). Figure 9.2 shows an example of how fan-out is determined in a simple logic circuit. Assuming that all of the logic gates in the system are standard logic devices, G1 must have

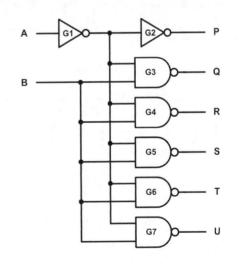

Figure 9.2 An example of determining the fan-out in a simple logic system. If all gates are standard logic devices, G1 must have a minimum fan-out of six because it drives inputs on G2, G3, G4, G5, G6, and G7

a minimum fan-out of six because it needs to drive inputs on six gates; G2, G3, G4, G5, G6, and G7.

9.2 Decoders

Decoders are used to convert information from one number system to another, such as binary to octal or binary to decimal. A simple two to four line decoder is shown in Figure 9.3. In this arrangement there are two inputs, A and B, and four outputs, Y_0, Y_1, Y_2 and Y_3.

The binary code appearing on A and B is decoded into one of the four possible states and corresponding output appears on the four output lines with Y_3 being the most significant. Because two to four and three to eight line decoders are frequently used as address decoders in computer systems (where memory and I/O devices are enabled by a low rather than a high state), the outputs are active-low, as indicated by the circles on the logic diagrams.

A basic two to four line decoder is shown in Figure 9.4, This arrangement uses two inverters and three two-input NAND gates. All four outputs are active-low; Y_0 will go low when A and B are both at logic 0, Y_1 will go low when A is at logic 0 and B is at logic 1, Y_2 will go low when A is at logic 1 and B is at logic 0, and Y_3 will go low when both A and B are at logic 1.

An improved version of the basic two to four line decoder is shown in Figure 9.5. The extra inverting gates are added at the inputs in order to reduce the fan-in to unity. On more complex arrangements (such as three to eight and four to sixteen decoders) these extra gates also help to avoid the situation where the inputs may present widely differing loads.

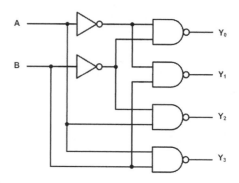

Figure 9.4 A basic two to four line decoder, see Figure 9.3(a)

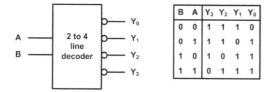

B	A	Y_3	Y_2	Y_1	Y_0
0	0	1	1	1	0
0	1	1	1	0	1
1	0	1	0	1	1
1	1	0	1	1	1

(a) Basic two to four line decoder

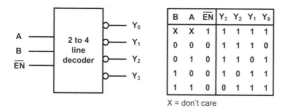

B	A	$\overline{EN}$	Y_3	Y_2	Y_1	Y_0
X	X	1	1	1	1	1
0	0	0	1	1	1	0
0	1	0	1	1	0	1
1	0	0	1	0	1	1
1	1	0	0	1	1	1

X = don't care

(b) Two to four line decoder with enable input

Figure 9.3 Two to four line decoders and their corresponding truth tables. The arrangement shown in (b) has a separate active-low enable input (when this input is high all of the outputs remain in the high state regardless of the A and B inputs)

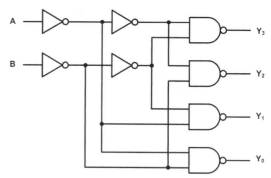

Figure 9.5 An improved two to four line decoder offering unity fan-in for both inputs, see Figure 9.3(b)

Figure 9.6 A 74LS139 dual two to four line decoder chip implemented using low-power Schottky TTL technology and supplied in a 16-pin DIP package

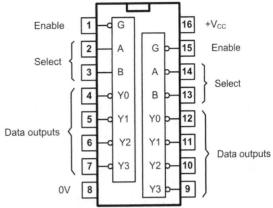

Figure 9.8 Pin connections for the 74LS139 dual two to four line decoder (viewed from above)

A typical dual two to four line MSI decoder chip, the 74LS139, is shown in Figure 9.6. The chip is implemented using low-power Schottky TTL technology. The device is supplied in a 16-pin DIP package (as shown in Figure 9.8) but is also available in ceramic and SOIC packaged versions (see page 99). The decoder is frequently used in logic circuits and address decoders.

A three to eight line decoder is shown in Figure 9.7. Note that, as with the two to four line

decoder shown in Figure 9.3(b), this arrangement has a separate active-low enable (EN) input. When this input is taken high, all eight of the decoder's outputs will go high. Only when the enable input is taken low will one (and only one) of the decoder's outputs go low. A typical example of a three to eight line decoder is the 74LS138. Like the 74LS139, this is a low-power Schottky device commonly used for address decoding.

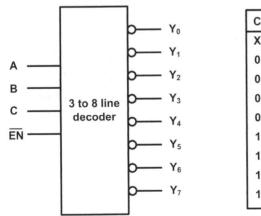

C	B	A	$\overline{EN}$	Y_7	Y_6	Y_5	Y_4	Y_3	Y_2	Y_1	Y_0
X	X	X	1	1	1	1	1	1	1	1	1
0	0	0	0	1	1	1	1	1	1	1	0
0	0	1	0	1	1	1	1	1	1	0	1
0	1	0	0	1	1	1	1	1	0	1	1
0	1	1	0	1	1	1	1	0	1	1	1
1	0	0	0	1	1	1	0	1	1	1	1
1	0	1	0	1	1	0	1	1	1	1	1
1	1	0	0	1	0	1	1	1	1	1	1
1	1	1	0	0	1	1	1	1	1	1	1

X = don't care

Figure 9.7 A three to eight line decoder and its corresponding truth table

9.3 Encoders

Encoders provide the reverse function to that of a decoder. In other words, they accept a number of inputs and then generate a binary code corresponding to the state of those inputs. Typical applications for encoders include generating a binary code corresponding to the state of a keyboard/keypad or generating BCD from a decade (ten-position) rotary switch.

A particularly useful form of encoder is one that can determine the priority of its inputs. This device is known as a **priority encoder** and its inputs are arranged in priority order, from lowest to highest priority. If more than one input becomes active, the input with the highest priority will be encoded and its binary coded value will appear on the outputs. The state of the other (lower priority) inputs will be ignored.

Figure 9.9 shows an eight to three line priority encoder together with its corresponding truth table. Note that this device has active-low inputs as well as active-low outputs. A practical realisation of this type of device, the 74LS148, is shown in Figure 9.10. Additional inputs and outputs (enable input, EI, and enable output, EO) are provided in order to permit **cascading** of

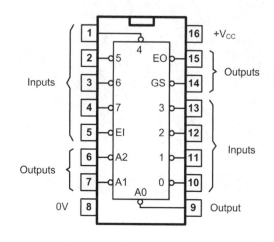

Figure 9.10 Pin connections for the 74LS148 eight to three line encoder (note that there is no need for a 0 input)

several devices of the same type. For example, a simple decimal keypad would require two such devices with the enable output of the first device connected to the enable input of the second device.

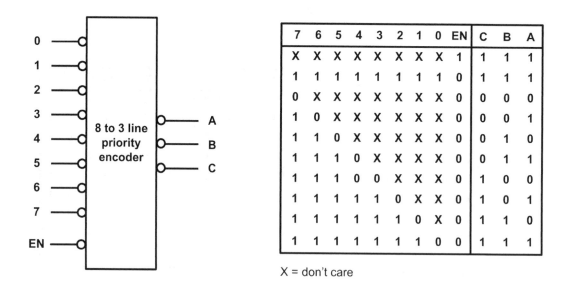

7	6	5	4	3	2	1	0	EN	C	B	A
X	X	X	X	X	X	X	X	1	1	1	1
1	1	1	1	1	1	1	1	0	1	1	1
0	X	X	X	X	X	X	X	0	0	0	0
1	0	X	X	X	X	X	X	0	0	0	1
1	1	0	X	X	X	X	X	0	0	1	0
1	1	1	0	X	X	X	X	0	0	1	1
1	1	1	0	0	X	X	X	0	1	0	0
1	1	1	1	1	0	X	X	0	1	0	1
1	1	1	1	1	1	0	X	0	1	1	0
1	1	1	1	1	1	1	0	0	1	1	1

X = don't care

Figure 9.9 An eight to three line priority encoder and its corresponding truth table. Note that the enable (EN) input is active-low

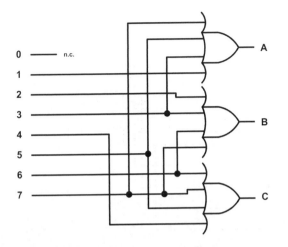

Figure 9.11 An octal to binary encoder

Test your understanding 9.1

Figure 9.11 shows the circuit of an octal to binary encoder. Verify the operation of this circuit by constructing its truth table.

Test your understanding 9.2

Redesign the octal to binary encoder shown in Figure 9.11 by adding inverters so that each of the input lines presents a fan-in of 1.

9.4 Multiplexers

Like encoders, multiplexers have several inputs. However, unlike encoders, they have only one output. Multiplexers provide a means of selecting data from one of several sources. Because of this, they are often referred to as **data selectors**.

Switch equivalent circuits of some common types of multiplexer are shown in Figure 9.12. The single two-way multiplexer in Figure 9.12(a) is equivalent to a simple SPDT (changeover) switch. The dual two-way multiplexer shown in Figure 9.12(b) performs the same function but two independent circuits are controlled from the same select signal. A single four-way multiplexer

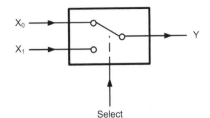

(a) Single two way

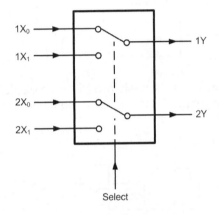

(b) Dual two way

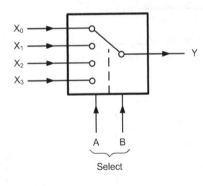

(c) Single four way

Figure 9.12 Switch equivalent circuits for some common types of multiplexer or data selector

is shown in Figure 9.12(c). Note that two digital select inputs are required, A and B, in order to place the switch in its four different states.

Block schematic symbols, truth tables and simplified logic circuits for two to one and four to one multiplexers are shown in Figures 9.13 to 9.16.

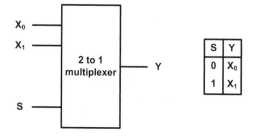

S	Y
0	X_0
1	X_1

Figure 9.13 A basic two to one multiplexer arrangement with its corresponding truth table

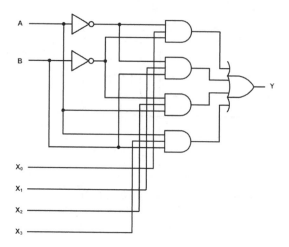

Figure 9.16 Logic gate arrangement for the four to one multiplexer

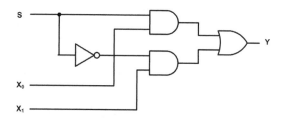

Figure 9.14 Logic circuit arrangement for the basic two to one multiplexer

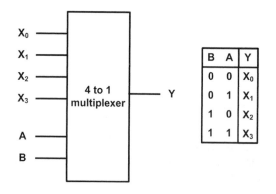

B	A	Y
0	0	X_0
0	1	X_1
1	0	X_2
1	1	X_3

Figure 9.15 A four to one multiplexer. The logic state of the A and B inputs determines which of the four logic inputs (X_0 to X_3) appears at the output

Test your understanding 9.3

Verify the operation of the two to one multiplexer circuit shown in Figure 9.14 by constructing its truth table.

Test your understanding 9.4

Verify the operation of the four to one multiplexer circuit shown in Figure 9.16 by constructing its truth table.

Test your understanding 9.5

Design a logic circuit arrangement that will function as an eight to one multiplexer. Confirm the operation of the circuit by constructing its truth table.

Test your understanding 9.6

Modify your answer to Test your understanding 9.5 so that each input has a fan-in of unity.

The practical realisation of an eight to one and 16 to one multiplexers, the 74LS151 and 74LS150 respectively, are shown in Figures 9.16 and 9.18. These MSI devices are both implemented using low-power Schottky TTL technology and they are supplied in either plastic DIP, ceramic or SOIC packages (see page 99).

A typical example of the use of multiplexers and decoders is shown in the simplified block schematic of the altimeter data selector shown in Figure 9.18. This arrangement uses a dual four channel multiplexer to select corresponding clock and data streams from the four ARINC 429 bus receivers.

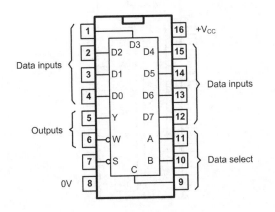

Figure 9.17 Pin connections for the 74LS151 eight input multiplexer (data selector)

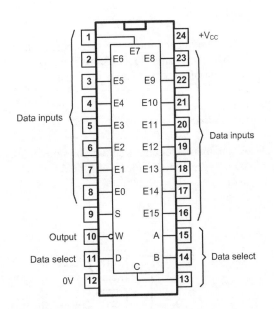

Figure 9.19 Pin connections for the 74LS150 sixteen input multiplexer (data selector)

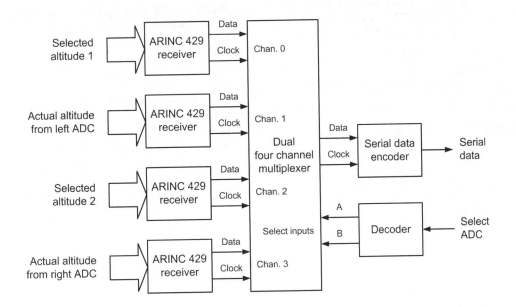

Figure 9.18 Four-channel altimeter data selector based on a dual four channel multiplexer

9.5 Multiple choice questions

1. A standard TTL logic gate has a fan-out of:
   (a) 1
   (b) 5
   (c) 10.

2. The equivalent number of standard input loads (of the same logic family) imposed by the input of a logic circuit is known as:
   (a) fan-in
   (b) fan-out
   (c) fan-load.

3. A simple logic circuit is shown in Figure 9.20. What is the minimum fan-out for G6?
   (a) 1
   (b) 5
   (c) 6.

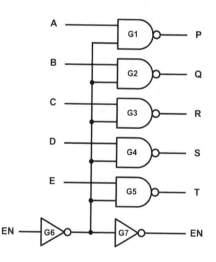

Figure 9.20 See Question 3

4. Another name for a multiplexer is:
   (a) a data selector
   (b) a bus transceiver
   (c) a shift register.

5. A four to one multiplexer has:
   (a) one select input
   (b) two select inputs
   (c) four select inputs.

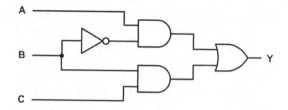

Figure 9.21 See Question 6

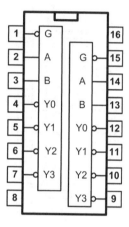

Figure 9.22 See Question 7

6. Which is the select input in the two way multiplexer shown in Figure 9.21?
   (a) A
   (b) B
   (c) C.

7. The integrated circuit device shown in Figure 9.22 is:
   (a) a dual two to four line decoder
   (b) a dual four to two line multiplexer
   (c) a dual four way data selector.

8. Additional enable inputs and outputs are provided in some encoders in order to permit:
   (a) inverting
   (b) cascading
   (c) error detection.

Chapter 10

Fibre optics

Optical fibres have been widely used as a transmission medium for ground-based long-haul data communications and in local area networks (LAN) for many years and they are now being introduced into the latest passenger aircraft to satisfy the need for wideband networked avionic and cabin entertainment systems.

By virtue of their light weight, compact size, and exceptionally wide bandwidth, optical fibres are ideally suited for use as a replacement for conventional copper network cabling. The technology is, however, relatively new in the civil aircraft industry and brings with it a whole new set of problems and challenges for those involved with aircraft operation and maintenance.

10.1 Advantages and disadvantages

Optical fibres offer some very significant advantages over conventional copper cables. These include:

- Optical fibres are lightweight and of small physical size
- Exceptionally wide bandwidth and very high data rates can be supported
- Relative freedom from electromagnetic interference
- Significantly reduced noise and cross-talk compared with conventional copper cables
- Relatively low values of attenuation within the medium
- High reliability coupled with long operational life
- Electrical isolation and freedom from earth/ground loops.

The reduction in weight that results from the use of fibre optical cabling can yield significant fuel savings. Copper cabling is typically five times heavier than polymer optical fibre cabling and 15 times heavier than silica optical fibre. On a large, latest generation aircraft with sophisticated

avionics, the total saving in weight can be as much as 1,300 kg.

There are very few disadvantages of optical fibres. They include:

- Industry resistance to the introduction of new technology
- Need for a high degree of precision when fitting cables and connectors
- Concerns about the mechanical strength of fibres and the need to ensure that cable bends have a sufficiently large radius to minimise losses and the possibility of damage to fibres.

10.2 Propagation in optical fibres

Essentially, an optical fibre consists of a cylindrical silica glass **core** surrounded by further glass **cladding**. The fibre acts as a channel (or waveguide) along which an electromagnetic wave can pass with very little loss.

Fibre optics is governed by the fundamental laws of reflection and refraction. For example, when a light wave passes from a medium of higher refractive index to one of lower refractive index, the wave is bent towards the normal, as shown in Figure 10.1(a). Conversely, when travelling from a medium of lower refractive index to one of higher refractive index, the wave will be bent away from the normal, as shown in Figure 10.1(b). In this latter case, some of the incident light will be reflected at the boundary of the two media and, as the angle of incidence is increased, the angle of refraction will also be increased until, at a critical value, the light wave will be totally reflected (i.e. the refracted ray will no longer exist, as shown in Fig. 10.2). The angle of incidence at which this occurs is known as the critical angle, θ_c. The value of θ_c depends upon the absolute refractive indices of the media and is given by:

$$\theta_c = \sqrt{\frac{2(n_1 - n_2)}{n_1}}$$

where n_1 and n_2 are the refractive indices of the more dense and less dense media respectively.

Optical fibres are manufactured by drawing silica glass from the molten state and they are

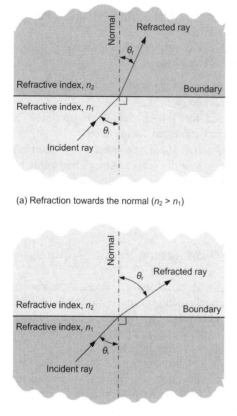

(a) Refraction towards the normal ($n_2 > n_1$)

(b) Refraction away from the normal ($n_2 < n_1$)

Figure 10.1 Refraction of a beam of light at a boundary

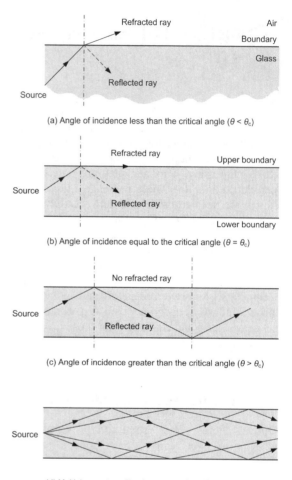

(a) Angle of incidence less than the critical angle ($\theta < \theta_c$)

(b) Angle of incidence equal to the critical angle ($\theta = \theta_c$)

(c) Angle of incidence greater than the critical angle ($\theta > \theta_c$)

(d) Multiple rays travelling by means of total internal reflection

Figure 10.2 Refraction and reflection at different angles of incidence

thus of cylindrical construction. The more dense medium (the **core**) is surrounded by the less dense medium (the **cladding**). Provided that the angle of incidence of the input wave is larger than the critical angle, the light wave will propagate inside the core by means of a series of **total internal reflections**. Any other light waves that are incident on the upper boundary at an angle $\theta_C > \theta$ will also propagate along the inner medium. Conversely, any light wave that is incident upon the upper boundary with $\theta < \theta_C$ will pass into the outer medium and there be lost by scattering and/or absorption.

10.2.1 Launching

Having briefly considered propagation within the

fibre, we shall turn our attention to the mechanism by which waves are launched into the fibre. The **cone of acceptance** (see Fig. 10.3) is the complete set of angles which will be subject to total internal reflection. Rays entering from the edges will take a longer path through the fibre but will travel faster because of the lower refractive index of the outer layer. The **numerical aperture** determines the bandwidth of the fibre and is given by:

Numerical aperture, $A = \sin \theta_a$

Clearly, when a number of light waves enter the system with differing angles of incidence, a number of waves (or modes) are able to

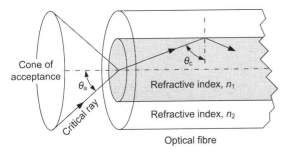

Figure 10.3 Cone of acceptance

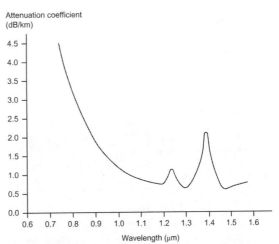

Figure 10.4 Attenuation in an optical fibre

propagate. This multimode propagation is relatively simple to achieve but has the attendant disadvantage that, since the light waves will take different times to pass through the fibre, the variation of transit time will result in dispersion which imposes an obvious restriction on the maximum bit rate that the system will support.

There are two methods for reducing multimode propagation. One uses a fibre of graded refractive index whilst the other uses a special single mode (or **monomode**) fibre. The inner core of this type of fibre is reduced in diameter so that it has the same order magnitude as the wavelength of the incident wave. This ensures that only one mode will successfully propagate.

10.2.2 Attenuation

The loss within an optical fibre arises from a number of causes including: **absorption**, **scattering** in the core (due to non-homogeneity of the refractive index), **scattering** at the core/cladding boundary, and losses due to **radiation** at bends in the fibre. Note that the **attenuation coefficient** of an optical fibre (see Fig. 10.4) refers only to losses in the fibre itself and neglects coupling and bending losses (which can be significant). In general, the attenuation of a good quality fibre can be expected to be less than 2 dB per km at a wavelength of 1.3 μm (**infra-red**). Hence a 50 m length of fibre can be expected to exhibit a loss of around 0.1 dB.

Whereas the attenuation coefficient of an optical fibre is largely dependent upon the quality and consistency of the glass used for the core and cladding, the attenuation of all optical fibres

varies widely with wavelength. The typical attenuation/wavelength characteristic for a monomode fibre is shown in Figure 10.4. It should be noted that the sharp peak at about 1.39 μm arises from excess absorption within the monomode fibre.

Monomode fibres are now a common feature of ground-based high speed data communication systems and manufacturing techniques have been developed that ensure consistent and reliable products with low attenuation and wide operational bandwidths. However, since monomode fibres are significantly smaller in diameter than their multimode predecessors (see Fig. 10.5), a consistent and reliable means of cutting, surface preparation, alignment and

Test your understanding 10.1

Define the term 'critical angle' in relation to a ray of light at a boundary between two optical media.

Test your understanding 10.2

Determine the critical angle for a glass-air interface (values of refractive index for glass and air are respectively 1.5 and 1.0).

interconnection is essential, and for this reason slower multimode fibres are still prevalent in current aircraft designs.

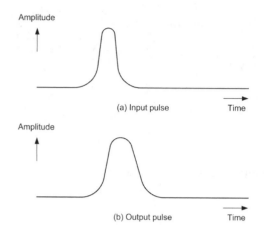

(a) Input pulse   Time

(b) Output pulse   Time

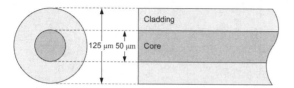

(a) Multimode fibre

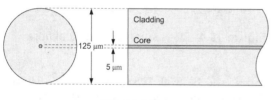

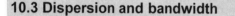

(b) Monomode fibre

Figure 10.5   Comparison of multimode and monomode fibres

Figure 10.7   Input and output pulses for Figure 10.6

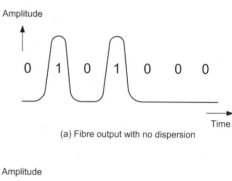

(a) Fibre output with no dispersion

10.3 Dispersion and bandwidth

A simple one-way (simplex) fibre optic data link is shown in Figure 10.6. The optical transmitter consists of an infra-red light emitting diode (LED) or low-power semiconductor laser diode coupled directly to the optical fibre. The diode is supplied with pulses of current from a bus interface. These pulses of current produce equivalent pulses of light that travel along the fibre until they reach the optical receiver unit. The optical receiver unit consists of a photodiode

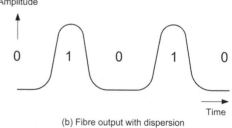

(b) Fibre output with dispersion

Figure 10.8   Effect of dispersion on the width of pulses received from an optical fibre

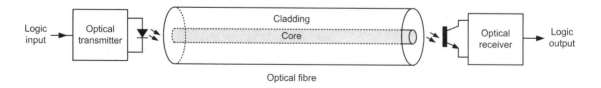

Optical fibre

Figure 10.6   A simple one-way fibre optic data link

or phototransistor that passes a relatively large current when illuminated and negligible current when not. The pulses of current at the transmitting end are thus replicated at the receiving end.

The maximum data rate (and consequently the bandwidth) of the optical data link depends on the ability of the system shown in Figure 10.6 to faithfully reproduce a train of narrow digital pulses. Unfortunately, in a multimode fibre different modes travel at different velocities, as shown earlier in Figure 10.2(d). This phenomenon is known as **dispersion** and it has the effect of stretching the output pulse, as shown in Figure 10.7(b).

When digital data is supplied to the optical transmitter, the stretching of pulses imposes an upper limit on the rate at which the pulses can be transmitted. In other words, the data rate is determined by the amount of dispersion simply because a longer bit interval means fewer bits can be transmitted in the same unit of time.

10.4 Practical optical networks

The Boeing 777 was the first commercial aircraft to enter production with an optical fibre based LAN for onboard data communications. The system was originally developed in the 1980s and it comprised an avionics local area network (AVLAN) fitted in the flight deck and electrical equipment bay together with a cabin network (CABLAN) fitted in the roof of the passenger cabin. These two fibre optical networks conform to the ARINC 636 standard which was adapted for avionics from the Fibre Distributed Interface (FDDI) in order to provide a network capable of supporting data rates of up to 100 Mbps.

10.4.1 Fibre optic cable construction

The construction of a typical fibre optic cable is shown in Figure 10.9. This comprises:

- Five optical fibres and two filler strands
- Separator tape
- Aramid yarn strength member
- An outer jacket.

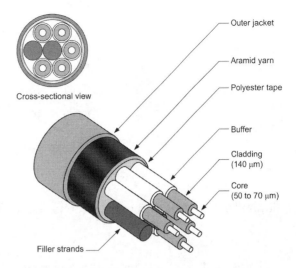

Figure 10.9 A typical fibre optic cable

The cable has an overall diameter of about 0.2 inches and the individual optical fibre strands have a diameter of 140 μm (approx. 0.0055 inches). A protective buffer covers each fibre and protects it during manufacture, increases mechanical strength and diameter in order to make handling and assembly easier. The buffers are coded in order to identify the fibres using colours (blue, red, green, yellow and white). The filler strands are made from polyester and are approximately 0.035 inches in diameter. A polyester separator tape covers the group of five fibres and two filler strands. This tape is manufactured from low-friction polyester and it serves to make the cable more flexible.

A layer of woven Aramid (or Kevlar) yarn provides added mechanical strength and protection for the cable assembly. The outer thermoplastic jacket (usually purple in colour) is fitted to prevent moisture ingress and also to provide insulation.

10.4.2 Fibre optic connectors

The essential requirements for connectors used with optical fibres are that they should be:

- Reliable
- Robust

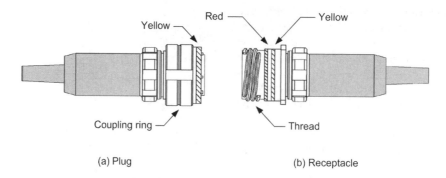

Yellow —   Red ——   — Yellow

Coupling ring —   — Thread

(a) Plug   (b) Receptacle

Figure 10.10 A typical fibre optic connector arrangement

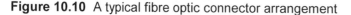

- Precise and repeatable (even after numerous mating operations)
- Suitable for installation without specialist tooling
- Low loss
- Low cost.

Whilst the loss exhibited by a connector may be quoted in absolute terms, it is often specified in terms of an equivalent length of optical fibre. If, for example, six connectors are used on a cable run and each connector has a loss of 0.5 dB the total connector loss will be 3 dB. This is equivalent to several kilometres of low-loss fibre!

A typical fibre optic cable connector arrangement is shown in Figure 10.10. This comprises:

- Alignment keys and grooves
- Guide pins and cavities
- Coloured alignment bands
- Three start threads.

Each connector has alignment keys on the plug and matching alignment grooves on the receptacle. These are used to accurately align the connector optical components; the guide pins in the plug fit into cavities in the receptacle when the plug and receptacle connect. In order to ensure that the connector is not over-tightened (which may cause damage to the fibres) the pins of the plug are designed to provide a buffer stop against the bottom of the cavities in the receptacle.

The plug and receptacle have ceramic contacts that are designed to make physical contact when

properly connected (the light signal passes through the through holes in the end of the ceramic contacts when they are in direct physical contact with each other).

The coupling nut on the plug barrel has a yellow band whilst the receptacle barrel has a red and a yellow band. A correct connection is made when the red band on the receptacle is at least 50 percent covered by the coupling nut. This position indicates an effective connection in which the optical fibres in the plug are aligned end-to-end with the fibre in the receptacle.

Three start threads on the plug and receptacle ensure a straight start when they join. The recessed receptacle components prevent damage from the plug if it strikes the receptacle at an angle. The plug and receptacle are automatically sealed in order to prevent the ingress of moisture and dust.

Key Point

Care must be taken when assembling optical fibre connectors both in terms of the correct alignment of the plugs and receptacles and cleanliness of the contact area (this is essential to ensure low loss and efficient coupling). Before attempting to examine the face or ceramic contacts of a connector arrangement it is essential to disconnect the cable from the equipment at both ends. The light from the optical fibre is invisible and can be intense enough to cause permanent damage to the eyes.

10.5 Optical network components

Several other components are found in optical fibre networks. These include **couplers** (with three or four ports), **switches** (using mirrors to deflect beams into different fibre strands), and **routers**. These last named devices are designed to control the routing of signals through the LAN and they comprise switches, processors, controllers, and one or more bus interfaces.

The router processor sends control signals to the **bypass switch unit** (BSU). Typical BSU control signals are:

- PRI HI
- PRI RTN
- SEC HI
- SEC RTN

A logic high on PRI HI or SEC HI connects the BSU to the fibre optic ring. A logic low on PRI HI or SEC HI disconnects the BSU from the fibre optic ring. The PRI RTN and SEC RTN control signals are grounds (active low inputs) to the BSU switch relays. The fibre optic interface changes BSU fibre optic signals to electronic signals and electronic signals to BSU fibre optic signals.

Test your understanding 10.3

Explain, with the aid of a sketch, how a beam of light propagates in an optical fibre.

Test your understanding 10.4

Explain the difference between multimode and monomode fibres.

Test your understanding 10.5

Explain the relationship between dispersion and bandwidth in an optical fibre.

Test your understanding 10.6

List THREE advantages of optical fibre data cables when compared with conventional copper cables.

Test your understanding 10.7

In relation to a multi-way optical fibre connector, explain the function and operation of:

(a) the keys and guides
(b) the guide pins
(c) the coloured bands.

10.6 Multiple choice questions

1. In an optical fibre, the refractive index of the core is:
   (a) the same as the cladding
   (b) larger than the cladding
   (c) smaller than the cladding.

2. A suitable optical transmitter for use with an optical fibre is:
   (a) a photodiode
   (b) a phototransistor
   (c) a low-power laser diode.

3. The typical core diameter for a monomode fibre is:
   (a) 5 μm
   (b) 50 μm
   (c) 125 μm.

4. In a multicore fibre optic cable:
   (a) the cores are colour coded
   (b) the cladding is colour coded
   (c) the buffers are colour coded.

5. The typical wavelength of light in a fibre is:
   (a) 1.3 μm
   (b) 13 μm
   (c) 130 μm.

6. What aircraft standard applies to fibre optic networks?
   (a) ARINC 429
   (b) ARINC 573
   (c) ARINC 636.

7. If a launcher has an acceptance angle of 45° the corresponding numerical aperture will be:
   (a) 0.5
   (b) 0.707
   (c) 1.

8. The typical cladding diameter for a monomode fibre is:
   (a) 5 μm
   (b) 50 μm
   (c) 125 μm.

9. If the numerical aperture of a fibre has a value of 1, this indicates that:
   (a) all of the light from the source is captured by the fibre
   (b) none of the light from a source is captured by a fibre
   (c) 50% of the light from a source is captured by a fibre.

10. When replacing a length of multi-way fibre optic cable, it is essential to ensure that:
    (a) the connectors are correctly aligned prior to mating the plug and the receptacle
    (b) none of the red band is covered when the coupling nut is tightened
    (c) all of the red band is covered when the coupling nut is tightened.

11. In order to support a high data rate, a fibre optic cable should have:
    (a) zero bandwidth
    (b) limited bandwidth
    (c) wide bandwidth.

12. A significant advantage of fibre optic networks in large passenger aircraft is:
    (a) lower installation costs
    (b) reduced weight
    (c) ease of maintenance.

13. A beam of light is directed towards a boundary between two optical media having different refractive indices. If the beam is incident at the critical angle the ray emerging from the boundary will travel:
    (a) away from the boundary
    (b) along the boundary
    (c) back towards the light source.

14. The bandwidth of an optical fibre is limited by:
    (a) attenuation and cross-talk in the cable.
    (b) modal dispersion occurring in the cable
    (c) the number and severity of bends in the cable.

15. Light propagates in a fibre optic by means of:
    (a) modal dispersion
    (b) continuous refraction
    (c) total internal reflection.

16. The attenuation of an optical fibre is typically:
    (a) less than 2 dB per km
    (b) between 2 dB and 20 dB per km
    (c) more than 20 dB per km.

17. The main advantage of monomode fibres is:
    (a) lower cost
    (b) smaller data cables
    (c) faster data rates.

18. A pulse of infrared light travelling down a multimode fibre optic becomes stretched. This is due to:
    (a) reflection
    (b) refraction
    (c) dispersion.

19. Attenuation in an optical fibre is due to:
    (a) absorption, dispersion, and radiation
    (b) absorption, scattering, and radiation
    (c) absorption, cross-talk, and noise.

20. Light waves in fibre optic cables are:
    (a) in the infrared spectrum
    (b) in the ultraviolet spectrum
    (c) in the visible spectrum.

Chapter 11 — Displays

Modern passenger aircraft employ a variety of different display technologies on the flight deck, including those based on conventional cathode ray tubes (CRT), light emitting diodes (LED) and liquid crystal displays (LCD). This chapter introduces these three main types of display and describes typical applications for each.

The trend is towards a uniform set of flight deck instruments using flat-panel displays but displaying information in formats that have evolved from earlier instruments and displays. Whilst there is a need to group instrument displays together in related functional areas (such as primary flight, navigation, and engine instruments), a high level of integration is now possible by combining data from different avionic systems and displaying it in different ways.

Flat-panel displays, such as active matrix liquid crystal displays (AMLCD), offer considerable savings in volume compared with CRT displays. Combined with developments in the miniaturization of electronic components, the use of modern surface mounted devices (SMD), and VLSI integrated circuits, this makes it possible to produce a complex multi-function instrument, complete with display, in a single enclosure. The single-box concept also helps to reduce the amount of cabling required and this, in turn, can simplify maintenance.

The latest AMLCD displays have performance parameter capabilities that exceed those of traditional CRT displays (see Fig. 11.1 and 11.2). The main advantages are in weight, power, volume (size), and reliability. However, AMLCD provide improved performance in several other areas including:

1. A high degree of uniformity of luminance, resolution and focus over the full display area
2. Ability to maintain display performance over a wide range of viewing angles
3. Immunity to ambient illumination washout and colour de-saturation
4. Ability to support a wide range of adjustable brightness levels
5. Ability to maximise the useable display area for a given panel size
6. A high degree of fault tolerance
7. Resistance to vibration and mechanical shock
8. Ability to maintain performance over a wide temperature range
9. Electromagnetic compatibility and ability to operate in the presence of high energy radiated RF fields (see Chapter 14).

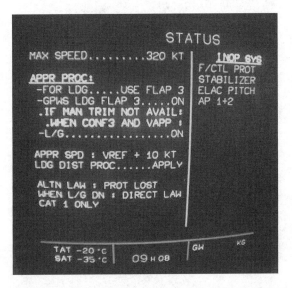

Figure 11.1 An example of information displayed on a flight deck CRT

Figure 11.2 Rear view of a CRT display showing external graphite coating and mumetal magnetic shielding

11.1 CRT displays

Apart from mechanical indicators, filament lamps, and moving coil meters, the cathode ray tube is the oldest display technology in current aircraft use.

Despite its age, the CRT offers a number of significant advantages, including the ability to provide an extremely bright colour display which can be viewed over a wide range of angles. For these two reasons, CRT displays are still found in modern aircraft despite the increasing trend to replace them with active matrix liquid crystal displays (AMLCD).

11.1.1 The cathode ray tube

The internal arrangement of a typical cathode ray tube is shown in Figure 11.3. The cathode, heater, grid and anode assembly forms an **electron gun** which produces a beam of electrons that is focused on the rear phosphor coating of the screen.

The heater raises the temperature of the cathode which is coated with thoriated tungsten (a material that readily emits electrons when heated). The negatively charged electrons form a cloud above the cathode (the electrons are literally 'boiled off' the cathode surface) and become attracted by the high positive potential that appears on the various anodes.

The flow of electrons is controlled by the grid. This structure consists of a fine wire mesh through which the electrons must pass. The grid is made negative with respect to the cathode and this negative potential has the effect of repelling the electrons. By controlling the grid potential it is possible to vary the amount of electrons passing through the grid thus controlling the

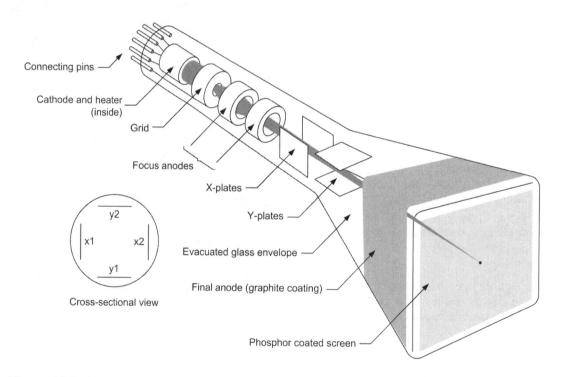

Figure 11.3 Internal arrangement of a CRT showing the path taken by the electron beam

intensity (or brightness) of the display on the screen.

The focus anodes consist of two or three tubular structures through which the electron beam passes. By varying the relative potential on these anodes it is possible to bend and focus the beam in much the same way as a light beam can be bent and focussed using a biconvex lens. The final anode consists of a graphite coating inside the CRT. This anode is given a very high positive potential (typically several kV) which has the effect of accelerating the beam of electrons as they travel towards it. The result is an electron beam of high energy impacting itself against the phosphor coating on the inside rear of the screen area. The energy liberated by the collision of the electrons with the phosphors is converted into light (the colour of the light depending on the particular colour of the phosphor at the point of impact).

11.1.2 Deflection

In order to move the beam of electrons to different parts of the screen (in other words, to be able to 'draw' on the screen) it is necessary to bend (or deflect) the beam. Two methods of deflection are possible depending on the size and application for the CRT. The method shown in Figure 11.4 uses **electrostatic deflection** (commonly used for small CRT displays). Using this method two sets of plates are introduced into the neck of the CRT between the focus anodes and the final anode. One pair of plates is aligned with the vertical plane (these X-plates provide deflection of the electron beam in the horizontal direction) whilst the other pair of plates is aligned in the horizontal plane (these Y-plates provide deflection of the electron beam in the vertical plane). By placing an electric charge (voltage) on the plates it is possible to bend the beam towards or away from a particular plate, as shown in Figure 11.4.

11.1.3 Scanning

In order to scan the full area of the CRT it is necessary to repeatedly scan the beam of

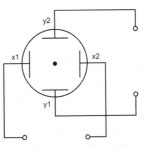

(a) No voltage applied to the deflection plates

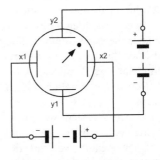

(b) y2 and x2 positive with respect to y1 and x1

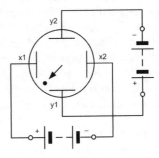

(c) y2 and x2 negative with respect to y1 and x1

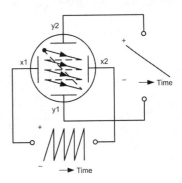

(d) Principle of raster scanning using a CRT

Figure 11.4   Deflecting the beam in a CRT using electrostatic deflection

electrons from top to bottom and left to right, as shown in Figure 11.4(d). The voltage waveforms required on the X and Y plates to produce the scanned **raster** must be ramp (sawtooth) shaped with different frequencies. For example, to produce the extremely crude four-line display shown in Figure 11.4(d) the ramp waveform applied to the X-plates would be 50 Hz whilst that applied to the Y-plates would be 200 Hz. A complete raster would then be scanned in a time interval of 20 ms (one fiftieth of a second).

A high resolution display will clearly require many more than just four lines however the principle remains the same. Suppose that we need to have 400 lines displayed and we are using a 100 Hz ramp for the Y-plates. The X-plates would then need to be supplied with a 40 kHz ramp waveform.

Having produced a raster, we can illuminate individual picture cells (**pixels**) by modulating the brightness of the beam (we can do this by applying a 'video' signal voltage to the cathode of the CRT. Essentially, we are then modulating the beam of electrons with the information that we need to display. In effect, the electron beam is being rapidly switched on and off in order to illuminate the individual pixels. Text can easily be displayed by this method by arranging characters into a **character cell** matrix. Typical arrangements of character cells are shown in Figure 11.5.

The alternative to electrostatic deflection is that of using an externally applied magnetic field to deflect the electron beam. This method is known as **electromagnetic deflection** and it is based on two sets of coils placed (externally) around the neck of the CRT. Comparable circuits of a CRT using electrostatic and electromagnetic deflection are shown in Figures 11.6 and 11.7 respectively.

Test your understanding 11.1

State TWO advantages and TWO disadvantages of CRT displays when compared with AMLCD displays.

Test your understanding 11.2

Refer to the circuit shown in Figure 11.6.

1. To which electrode of the CRT is the video (intensity modulation) signal applied?

2. What approximate voltage appears on the first anode of the CRT?

3. What is the CRT heater voltage?

4. What is the approximate range of voltages applied to the third anode?

5. What is the function of TR21 and TR22?

Key Point

CRT displays use an electron gun assembly to produce an accurately focussed beam of electrons which then impacts against a phosphor coated screen. By controlling the amount of electrons (i.e. modulating the current in the beam) it is possible to control the intensity of the light produced.

Key Point

In order to cover the full screen area of a CRT display it is necessary to scan the beam (up and down and left to right) in order to create a raster. The raster is generated by applying ramp waveforms of appropriate frequency to the X and Y-deflection system.

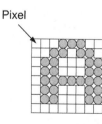

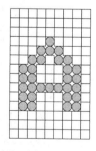

(a) 8 × 8 character cell   (b) 9 ×14 character cell

Figure 11.5 Examples of character cells

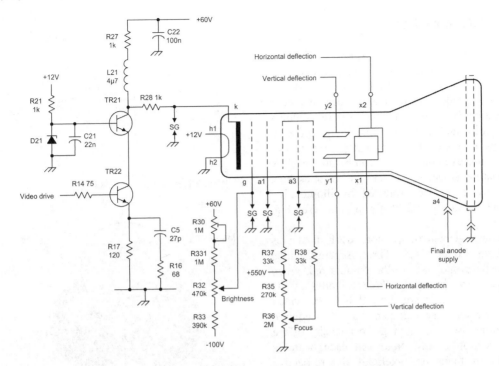

Figure 11.6 CRT circuitry using electrostatic deflection (SG = spark gap)

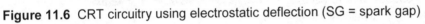

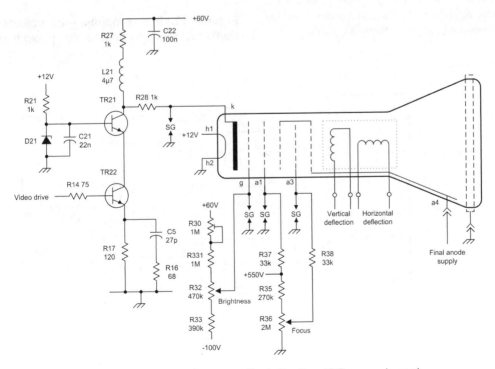

Figure 11.7 CRT circuitry using electromagnetic deflection (SG = spark gap)

11.1.4 Colour displays

By introducing a pattern of phosphors of different colours and by using a more complicated CRT with three different cathodes (see Fig. 11.9) it is possible to produce a CRT that can display colour information. By combining three different colours (red, green and blue phosphors) in different amounts it is possible to generate a range of colours. For example, yellow can be produced by illuminating adjacent red and green phosphors whilst white can be produced by illuminating adjacent red, green and blue phosphors (see Fig. 11.10).

The arrangement of a colour CRT display is shown in Figure 11.8. Three separate video signals (corresponding to the colours red, green and blue) are fed to the three cathodes of the CRT. These signals are derived from the video processing circuitry that generates the required waveforms used for varying the intensity of the three electron beams. Note that each beam is brought to focus on pixels of the respective colour (for example, the beam generated by the red cathode only coincides with the red phosphors). A synchronising system generates the scanning ramp waveforms and ensures that the time relationship between them is correct.

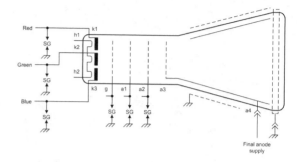

Figure 11.9 A colour CRT

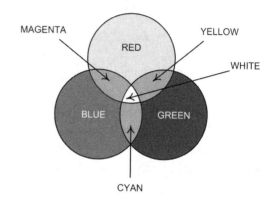

Figure 11.10 Generation of colours by illuminating red, green and blue phosphors

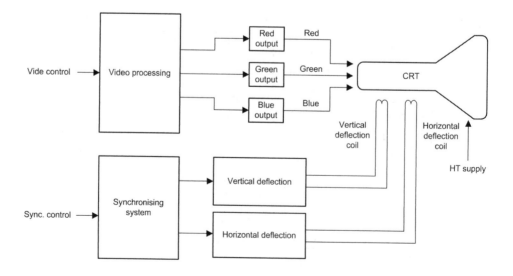

Figure 11.8 Arrangement of a colour display

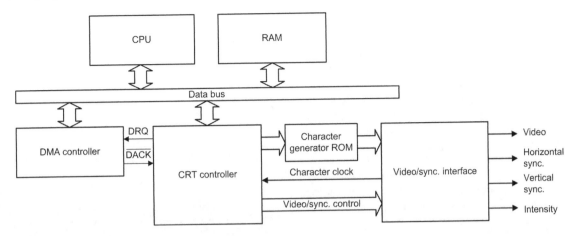

Figure 11.11 Typical CRT controller arrangement

11.1.5 CRT control

A dedicated CRT controller integrated circuit acting in conjunction with a video/synchronising interface provides the necessary control signals for the CRT (including signals that are used to synchronise the X and Y-scanning ramp waveforms). In turn, the CRT controller acts under the control of a dedicated CPU which accepts data from the bus and buffers the data ready for display. Direct Memory Access (DMA) is used to minimise the burden on the CPU (which would otherwise need to process data on an individual byte or word basis).

The patterns required to generate the displayed characters (see Fig. 11.5) are stored in a dedicated character generator ROM (see Fig. 11.11). Data for each scan line is read out from this ROM and assembled into a serial stream of bits which are fed to the appropriate video signal channels.

11.2 Light emitting diodes (LED)

Light emitting diodes (LED) can be used as general-purpose indicators. When compared with conventional filament lamps they operate from significantly smaller voltages and currents. LEDs

Table 11.1 Characteristics of various types of LED

Parameter	Type of LED			
	Miniature	Standard	High efficiency	High intensity
Diameter (mm)	3	5	5	5
Max. forward current (mA)	40	30	30	30
Typical forward current (mA)	12	10	7	10
Typical forward voltage drop (V)	2.1	2.0	1.8	2.2
Max. reverse voltage (V)	5	3	5	5
Max. power dissipation (mW)	150	100	27	135
Peak wavelength (nm)	690	635	635	635

are also very much more reliable than filament lamps. Most LEDs will provide a reasonable level of light output when a forward current of between 5 mA and 20 mA is applied. Light emitting diodes are available in various formats with the round types being most popular. Round LEDs are commonly available in the 3 mm and 5 mm (0.2 inch) diameter plastic packages and also in a 5 mm × 2 mm rectangular format. The viewing angle for round LEDs tends to be in the region of 20° to 40°, whereas for rectangular types this is increased to around 100°. Table 11.1 shows the characteristics of some common types of LED.

11.2.1 Spectral response

Light of different colours can be produced by using different semiconductor materials in the construction of an LED. However, there is a wide variation in both the efficiency and light output of LED of different colours. For this reason, red displays tend to be most common (with a peak output at around 650 nm). Note that this is towards one end of the visible spectrum, as shown in Figure 11.12.

11.2.3 Seven segment displays

LED displays are frequently used to display numerical data. The basis of such displays is the seven segment indicator (see Fig. 11.13) which is often used in groups of between three and five digits to form a complete display. The arrangement of the individual segments of a seven segment indicator is shown in Figure 11.14. The segments are distinguished by the letters, a to g. Since each segment comprises an individual LED it is necessary to use logic to decode binary (or binary coded decimal) data in order to illuminate the correct combination of segments to display a particular digit. For example, the number '1' can be displayed by simultaneously illuminating segments b and c whilst the number '2' requires that segments a, b, g, e, and d should be illuminated. The circuit of a seven segment display is shown in Figure 11.15 whilst a typical decoder and decoder truth table are shown in Figures 11.16 and 11.17 respectively.

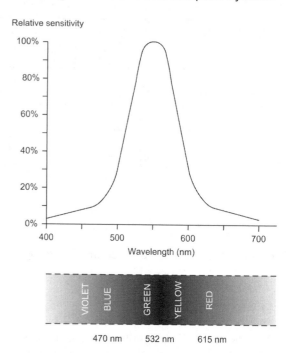

Figure 11.12 Typical spectral response for the human eye

Figure 11.13 Typical example of a four-digit seven segment display

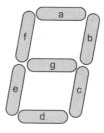

Figure 11.14 Segment identification within a seven segment display

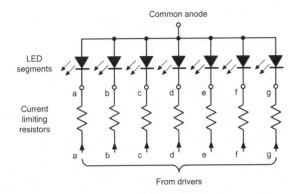

Figure 11.15 Circuit of a seven segment display (including series current limiting resistors)

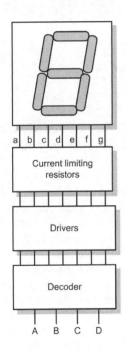

Figure 11.16 Seven segment decoder/driver

A B C D	Display
0 0 0 0	0
0 0 0 1	1
0 0 1 0	2
0 0 1 1	3
0 1 0 0	4
0 1 0 1	5
0 1 1 0	6
0 1 1 1	7
1 0 0 0	8
1 0 0 1	9
1 0 1 0	c
1 0 1 1	c
1 1 0 0	u
1 1 0 1	c
1 1 1 0	t
1 1 1 1	

Figure 11.17 Truth table for the decoded seven segment display

Test your understanding 11.3

State TWO advantages of LED indicators when compared with conventional filament lamp indicators.

Test your understanding 11.4

Explain the function of a seven segment decoder and illustrate your answer with a sketch showing a typical decoder arrangement.

11.3 Liquid crystal displays (LCD)

Liquid crystals have properties that can be considered to be somewhere between those of a solid and those of a liquid. Solids have a rigid molecular structure whilst the molecules in liquids change their orientation and are able to move. A particular property of liquid crystals that makes them attractive for use as the basis of electronic displays is that the orientation of molecules (and consequently the passage of light through the crystal) can be controlled by the application of an electric field.

11.3.1 Types of LCD

LCD displays can be either reflective or backlit according to whether the display uses incident light or contains its own light source. Figure 11.18 shows the construction of both types of display. Note that, unlike LED, liquid crystal displays emit no light of their own and, as a consequence, they need a light source in order to operate.

Larger displays can be easily made that combine several digits into a single display. This makes it possible to have integrated displays where several sets of information are shown on a common display panel. Figure 11.19 shows the comparison of typical LCD and LED aircraft displays that show the same information. Note that each LCD display replaces several seven segment LED displays.

Figure 11.20 shows an example of the use of three-digit seven segment displays for battery bus voltage indication whilst Figure 11.21 shows the method of interfacing a three-digit LCD to a microcontroller.

Figure 11.22 shows a larger alphanumeric LCD which is organised on the basis of 40 characters arranged in two lines. Displays of this type are ideal for showing short text messages.

11.3.2 Passive matrix displays

In order to display more detail (for example, text and graphics characters) LCD displays can be built using a matrix of rows and columns in order

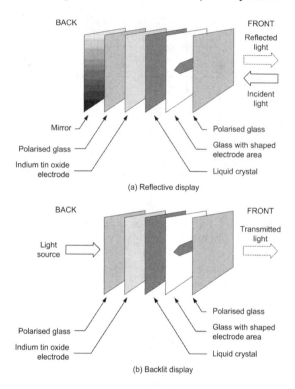

Figure 11.18 An enlarged section of an LCD showing a single character segment

to produce a display that consists of a rectangular matrix of cells. The electrodes used in this type of display consist of rows and columns of horizontal and vertical conductors respectively. The rows and columns can be separately addressed (in a similar manner to that used for a memory cell matrix, see page 67) and individual display cells can thus be illuminated. Passive matrix displays have a number of disadvantages, notably that they have a relatively slow response time and the fact that the display is not as sharp (in terms of resolution) as that which can be obtained from an active matrix display. A typical passive matrix display is shown in Figure 11.21.

11.3.3 Active matrix displays

Active matrix LCD (AMLCD) use thin film transistors (TFT) fabricated on a glass substrate to that they are an integral part of a display. Each transistor acts as a switch that transfers charge to

(a) LCD display

(b) LED display

Figure 11.19 Alternative standby engine indicators based on (a) LCD and (b) LED displays

Figure 11.20 Three-digit seven segment displays used for battery bus voltage indication

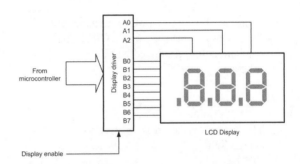

Figure 11.21 Method of interfacing a three-digit LCD to a microcontroller

Figure 11.22 A 40 character two line alphanumeric display using passive LCD technology

an individual display element. The transistors are addressed on a row/column basis as with the passive matrix display. By controlling the switching, it is possible to transfer precise amounts of charge into the display and thus exert a wide range of control over the light that is transmitted through it.

Colour AMLCD comprise a matrix of pixels that correspond to three colours; red, green and blue. By precise application of charges to the appropriate pixels it is possible to produce displays that have 256 shades of red, green and blue (making a total of more than 16 million colours). High resolution colour AMLCD make it possible to have aircraft displays with a full graphics capability.

Test your understanding 11.5

Explain the difference between active and passive matrix LCD. State TWO advantages of active matrix displays.

11.4 Multiple choice questions

1. The peak wavelength of light emitted from a red LED is typically:
   (a) 475 nm
   (b) 525 nm
   (c) 635 nm.

2. The spectral response of the human eye peaks at around:
   (a) 450 nm
   (b) 550 nm
   (c) 650 nm.

3. Adjacent phosphors of blue and green are illuminated on the screen of a CRT. The resultant colour produced will be:
   (a) cyan
   (b) white
   (c) magenta.

4. The human eye is most sensitive to:
   (a) red light
   (b) blue light
   (c) green light.

5. A seven segment LED display has segments a, b, c, d, and g illuminated. The character displayed will be:
   (a) 2
   (b) 3
   (c) 5.

6. The BCD input to a seven segment display decoder is 1001. The digit displayed will be:
   (a) 7
   (b) 8
   (c) 9.

7. The potential at the gird of a CRT is:
   (a) the same as the cathode
   (b) negative with respect to the cathode
   (c) positive with respect to the cathode.

8. The function of the final anode of a CRT is:
   (a) accelerating the electron beam
   (b) deflecting the electron beam
   (c) controlling the brightness of the display.

9. Electromagnetic deflection of a CRT uses:
   (a) coils inside the CRT
   (b) X- and Y-plates inside the CRT
   (c) coils around the neck of the CRT.

10. The three beams in a colour CRT are associated with the colours:
    (a) red, yellow and blue
    (b) red, green and blue
    (c) green, blue and yellow.

11. When compared with CRT displays, AMLCD displays have:
    (a) larger volume and lower reliability
    (b) larger volume and greater reliability
    (c) smaller volume and greater reliability.

12. The final anode of a CRT display requires:
    (a) a low voltage AC supply
    (b) a high voltage AC supply
    (c) a high voltage DC supply.

13. The deflecting waveform supplied to the plates of an electrostatic CRT will be:
    (a) a sine wave
    (b) a ramp wave
    (c) a square wave.

14. The typical value of maximum forward current for an LED indicator is:
    (a) 0.03 A
    (b) 0.3 A
    (c) 3 A.

15. The patterns required to display alphanumeric characters in a CRT controller are stored in:
    (a) static RAM
    (b) dynamic RAM
    (c) character generator ROM.

16. When compared with passive matrix LCD, AMLCD are:
    (a) faster and sharper
    (b) faster and less sharp
    (c) slower and less sharp.

Chapter 12 ESD

Earlier we said that advancements in technology were bringing new challenges for those involved with the operation and maintenance of modern passenger aircraft. One of those challenges is associated with the handling of semiconductor devices that are susceptible to damage from stray electric charges. This is a problem that can potentially affect a wide range of electronic equipment fitted in an aircraft (see Fig. 12.1) and can have wide ranging effects, including total failure of the LRU but without any visible signs of damage!

Electrostatic Sensitive Devices (ESD) are electronic components and other parts that are prone to damage from stray electric charge. This problem is particularly prevalent with modern LSI and VLSI devices but it also affects other components such as metal oxide semiconductor (MOS) transistors, microwave diodes, displays, and many other modern electronic devices.

Extensive (and permanent) damage to static sensitive devices can result from mishandling and inappropriate methods of storage and transportation. This chapter provides background information and specific guidance on the correct handling of ESD.

Figure 12.1 Part of the avionics bay of a modern passenger aircraft containing LRUs which use large numbers of electrostatic sensitive devices (ESD)

12.1 Static electricity

Static electricity is something that we should all be familiar with in its most awesome manifestation, lightning (see Fig. 12.2). Another example of static electricity that you might have encountered is the electric shock received when stepping out of a car. The synthetic materials used for clothing as well as the vehicle's interior are capable of producing large amounts of static charge which is only released when the hapless driver or passenger sets foot on the ground!

When two dissimilar, initially uncharged non-conducting materials are rubbed together, the friction is instrumental in transferring charge from one material to the other and consequently

Figure 12.2 Lightning (a natural example of static electricity) results from the build up of huge amounts of static charge

raising the electric potential that exists between them.

12.1.1 The triboelectric series

The triboelectric series classifies different materials according to how well they create static electricity when rubbed with another material. The series is arranged on a scale of increasingly positive and increasingly negative materials.

The following materials give up electrons and become positive when charged (and so appear as positive on the **triboelectric scale**) when rubbed against other materials:

* Air (most positive)
* Dry human skin
* Leather
* Rabbit fur
* Glass
* Human hair
* Nylon
* Wool
* Lead
* Cat fur
* Silk
* Aluminium
* Paper (least positive).

The following are examples of materials that do not tend to readily attract or give up electrons when brought in contact or rubbed with other materials (they are thus said to be neutral on the triboelectric scale):

* Cotton
* Steel

The following materials tend to attract electrons when rubbed against other materials and become negative when charged (and so appear as negative on the triboelectric scale):

* Wood (least negative)
* Amber
* Hard rubber
* Nickel, copper, brass and silver
* Gold and platinum
* Polyester
* Polystyrene

* Saran
* Polyurethane
* Polyethylene
* Polypropylene
* Polyvinylchloride (PVC)
* Silicon
* Teflon (most negative).

The largest amounts of induced charge will result from materials being rubbed together that are at the extreme ends of the triboelectric scale. For example, PVC rubbed against glass or polyester rubbed against dry human skin. Note that a common complaint from people working in a dry atmosphere is that they produce sparks when touching metal objects. This is because they have dry skin, which can become highly positive in charge, especially when the clothes they wear are made of man-made material (such as polyester) which can easily acquire a negative charge. The effect is much less pronounced in a humid atmosphere where the stray charge can leak away harmlessly into the atmosphere. People that build up static charges due to dry skin are advised to wear all-cotton clothes (recall that cotton is neutral on the triboelectric scale). Also, moist skin tends to dissipate charge more readily.

Human hair becomes positive in charge when combed. A plastic comb will collect negative charges on its surface. Since similar charges repel, the hair strands will push away from each other, especially if the hair is very dry. The comb (which is negatively charged) will attract objects with a positive charge (like hair). It will also attract material with no charge, such as small pieces of paper. You will probably recall demonstrations of this effect when you were studying science at school.

Electric charge can also be produced when materials with the same triboelectric polarity are rubbed together. For example, rubbing a glass rod with a silk cloth will charge the glass with positive charges. The silk does not retain any charges for long. When both of the materials are from the positive side of the triboelectric scale (as in this example) the material with the greatest ability to generate charge will become positive in charge. Similarly, when two materials that are both from the negative end of the triboelectric scale are rubbed together, the one with the greatest tendency to attract charge will become negative in charge.

Representative values of electrostatic voltages generated in some typical working situations are shown in Table 12.1. Note the significant difference in voltage generated at different values of relative humidity.

Key Point

Very large electrostatic potentials can be easily generated when different materials are rubbed together. The effect is much more pronounced when the air is dry.

Test your understanding 12.1

Explain, in relation to electric charge, what happens when a glass rod is rubbed with a polyester cloth.

Test your understanding 12.2

Explain the importance of the triboelectric series. Give ONE example of a material from the positive end of the triboelectric scale and ONE example of a material from the negative end of the triboelectric scale.

12.2 Static sensitive devices

All modern microelectronic components are prone to damage from stray electric charges but some devices are more prone to damage than others. Devices that are most prone to damage tend to be those that are based on the use of field effect technology rather than bipolar junction technology. They include CMOS logic devices (such as logic gates and MSI logic), MOSFET devices (such as transistors), NMOS and PMOS VLSI circuits (used for dynamic memory devices, microprocessors, etc). Microwave transistors and diodes (by virtue of their very small size and junction area) are also particularly static sensitive as are some optoelectronic and display devices. If in doubt, the moral here is to treat *any* semiconductor device with great care and to *always* avoid situations in which stray static charges may come into contact with a device.

Printed circuit board assemblies can also be prone to damage from electrostatic discharge. In general, printed circuit board mounted components are at less risk than individual semiconductors. The reason for this is that the conductive paths that exist in a printed circuit can often help to dissipate excessive static charges that might otherwise damage un-mounted semiconductor devices (there are no static dissipative paths when a transistor, diode or integrated circuit is handled on its own).

Table 12.2 provides a guide as to the relative susceptibility of various types of semiconductor device to damage from static voltages.

Table 12.1 Representative values of electrostatic voltages generated in typical work situations

Situation	Typical electrostatic voltage generated	
	20% relative humidity	80% relative humidity
Walking over a wool/nylon carpet	35 kV	1.5 kV
Sliding a plastic box across a carpet	18 kV	1.2 kV
Removing parts from a polystyrene bag	15 kV	1 kV
Walking over vinyl flooring	11 kV	350 V
Removing shrink wrap packaging	10 kV	250 V
Working at a bench wearing overalls	8 kV	150 V

Table 12.2 Representative values of static voltage susceptibility for different types of semiconductor

Type of device	Typical static voltage susceptibility
CMOS logic	250 V to 1 kV
TTL logic	550 V to 2.5 kV
Bipolar junction transistors	150 V to 5 kV
Dynamic memories	20 V to 100 V
VLSI microprocessor	20 V to 100 V
MOSFET transistors	50 V to 350 V
Thin film resistors	300 V to 3 kV
Silicon controlled rectifiers	4 kV to 15 kV

12.3 ESD warnings

Static sensitive components (including printed circuit board cards, circuit modules, and plug-in devices) are invariably marked with warning notices. These are usually printed with black text on yellow backgrounds, as shown in Figures 12.3 to 12.7.

Figure 12.3 A typical ESD warning label

Figure 12.4 ESD warning notice (third from bottom) in the avionics bay of a Boeing 737

12.4 Handling and transporting ESD

Special precautions must be taken when handling, transporting, fitting and removing ESD. These include the following:

1. Use of wrist straps which must be worn when handling ESD. These are conductive bands that are connected to an effective ground point by means of a short wire lead. The lead is usually fitted with an integral 1 MΩ resistor which helps to minimise any potential shock hazard to the wearer (the series resistor serves to limit the current passing through the wearer in the event that he/she may come into contact with a live conductor). Wrist straps are usually stored at strategic points on the aircraft (see Figure 12.5) or may be carried by maintenance

technicians. Figure 12.6 shows a typical wrist strap being used for a bench operation whilst Figures 12.7 and 12.8 show ESD warning notices associated with the wearing of wrist straps

2. Use of heel straps which work in a similar manner to wrist straps
3. Use of static dissipative floor and bench mats
4. Avoidance of very dry environments (or at least the need to take additional precautions when the relative humidity is low)

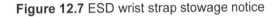

Figure 12.7 ESD wrist strap stowage notice

Figure 12.5 Typical on-board stowage for a wrist strap

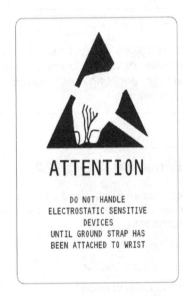

Figure 12.8 ESD wrist strap warning notice

5. Availability of ground jacks (see Fig. 12.6)
6. Use of grounded test equipment
7. Use of low-voltage soldering equipment and anti-static soldering stations (low-voltage soldering irons with grounded bits)
8. Use of anti-static insertion and removal tools for integrated circuits
9. Avoidance of nearby high-voltage sources (e.g. fluorescent light units)
10. Use of anti-static packaging (static sensitive components and printed circuit boards should be stored in their anti-static packaging until such time as they are required for use).

Note that there are three main classes of materials used for protecting static sensitive devices. These are **conductive materials** (such as

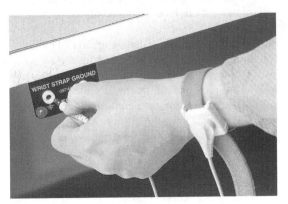

Figure 12.6 Using a wrist strap for a bench operation (note the grounding jack connector)

metal foils, and carbon impregnated synthetic materials), **static dissipative materials** (a cheaper form of conductive material), and so-called **anti-static materials** (these are materials that are neutral on the triboelectric scale, such as cardboard, cotton, and wood). Of these, conductive materials offer the greatest protection whilst anti-static materials offer the least protection.

Key Point

Stray static charges can very easily damage static-sensitive devices. Damage can be prevented by adopting the correct ESD handling procedures.

Test your understanding 12.3

Which one of the following semiconductor devices is likely to be most susceptible to damage from stray static charges?

1. A dynamic memory

2. A silicon controlled rectifier

3. A bipolar junction transistor.

Test your understanding 12.4

Which one of the following situations is likely to produce the greatest amount of stray static charge?

1. Removing a PVC shrink wrap on a dry day

2. Walking on a vinyl floor on a wet day

3. Sitting at a bench wearing a wrist strap.

Test your understanding 12.5

Explain the difference between conductive and static dissipative materials for ESD protection.

12.5 Multiple choice questions

1. A particular problem with the build-up of static charge is that:
   (a) it is worse when wet
   (b) it is invariably lethal
   (c) it cannot easily be detected.

2. The typical resistance of a wrist strap lead is:
   (a) 1 Ω
   (b) 1 kΩ
   (c) 1 MΩ.

3. Which one of the following devices is most susceptible to damage from stray static charges:
   (a) a power rectifier
   (b) a TTL logic gate
   (c) a MOSFET transistor.

4. The static voltage generated when a person walks across a carpet can be:
   (a) no more than about 10 kV
   (b) between 10 kV and 20 kV
   (c) more than 20 kV.

5. Which of the materials listed is negative on the triboelectric scale?
   (a) glass
   (b) silk
   (c) polyester.

6. When transporting ESD it is important to:
   (a) keep them in a conductive package
   (b) remove them and place them in metal foil
   (c) place them in an insulated plastic package.

7. To reduce the risk of damaging an ESD during soldering it is important to:
   (a) use only a low-voltage soldering iron
   (b) use only a mains operated soldering iron
   (c) use only a low-temperature soldering iron.

8. Which one of the following items of clothing is most likely to cause static problems?
   (a) nylon overalls
   (b) a cotton T-shirt
   (c) polyester-cotton trousers.

Chapter 13   Software

Aircraft software is something that you can't see and you can't touch yet it must be treated with the same care and consideration as any other aircraft part. This chapter deals with aspects of the maintenance of the safety-critical software found in modern aircraft.

At the outset it is important to realise that software encompasses both the executable code (i.e. the programs) run on aircraft computers as well as the data that these programs use. The term also covers the operating systems (i.e. system software) embedded in aircraft computer systems. All of these software parts require periodic upgrading as well as modification to rectify problems and faults that may arise as a result of operational experience (see Fig. 13.1 for an example emphasising the importance of this).

The consequences of software failure can range from insignificant (no effect on aircraft performance) to catastrophic (e.g. major avionic system failure, engine faults, etc). Because of this it is important that you should have an understanding of the importance of following correct procedures for software modification and

ATA 73—FUEL AND CONTROL
EEC SOFTWARE—MODIFICATION

There have been six events of in-service loss of engine parameters displayed in the aircraft cockpit, combined with freezing of engine power setting for the affected engine. Subsequent investigation has established the cause to be a fault in engine software. A safety assessment has identified that this could result in hazardous asymmetric thrust, if the event were to occur during a take-off roll and in response the crew attempted to abort the take-off. This Airworthiness Directive requires the introduction of revised engine software.

Figure 13.1 An extract from a recent Airworthiness Directive identifying the need for software modification

upgrading. Once again, this is an area of rapidly evolving technology which brings with it many new challenges.

13.1 Software classification

Aircraft software can be divided into five levels according to the likely consequences of its failure, as shown in Table 13.1. The highest level of criticality (Level A) is that which would have

Table 13.1 Levels of failure

Level	Type of failure	Failure description	Probability	Likelihood of failure (per flight hour)
A	Catastrophic failure	Aircraft loss and/or fatalities	Extremely improbable	Less than 10^{-9}
B	Hazardous/severe major failure	Flight crew cannot perform their tasks; serious or fatal injuries to some occupants	Extremely remote	Between 10^{-7} and 10^{-9}
C	Major failure	Workload impairs flight crew efficiency; occupant discomfort including injuries	Remote	Between 10^{-5} and 10^{-7}
D	Minor failure	Workload within flight crew capabilities; some inconvenience to occupants	Probable	Greater than 10^{-5}
E	No effect	No effect	Not applicable	

catastrophic consequences whilst the lowest level of criticality is that which would have no significant impact on the operation of the aircraft. In between these levels the degree of criticality is expressed in terms of the additional workload imposed on the flight crew and, in particular, the ability of the flight crew to manage the aircraft without having access to the automatic control /or flight information that would have otherwise been provided by the failed software. Table 13.2 provides examples of software applications and level of software criticality associated with each.

13.2 Software certification

The initial certification of an aircraft requires that the Design Organisation (DO) shall provide evidence that the software has been designed, tested and integrated with the associated hardware in a manner that satisfies standard DO-178B/ED-12B (or an agreed equivalent standard). In order to provide an effective means of software identification and change control, a software configuration management plan (CMP) (e.g. as defined in Part 7 of DO-178B/ED-12B) is required to be effective throughout the life of the equipment (the CMP must be devised and maintained by the relevant DO).

Post-certificate modification of equipment in the catastrophic, hazardous, or major categories (Levels A, B, C and D, see Tables 13.1 and 13.2) must not be made unless first approved by the DO. Hence all software upgrades and modifications are subject to the same approval procedures as are applied to hardware modifications. This is an important point that recognises the importance of software as an 'aircraft part'. Any modifications made to software must be identified and controlled in accordance with the CMP. Guidance material is provided DO-178B.

Test your understanding 13.1

Explain the purpose of a software Configuration Management Plan and specify the standard that applies to the development and management of safety critical aircraft software.

Table 13.2 Examples of software levels

Level	Typical aircraft applications (see Appendix 1 for acronyms)
A	AHRS GPS/ILS/MLS/FLS SATNAV VOR ADF
B	TCAS ADSB Transponder Flight Displays
C	DME VHF voice communications
D	AHRS Automatic Levelling CMC/CFDIU Data Loader Weather Radar
E	In-flight entertainment

Key Point

Aircraft software can be taken to encompass both the executable code (i.e. programs) as well as the data that these programs use. The term also covers the operating systems (i.e. system software) embedded in aircraft computer systems. All of these software parts require periodic upgrading as well as modification to rectify problems and faults that may arise.

Key Point

DO-178B is the de-facto standard for the development and management of safety critical software used in aviation. DO-178B certification deals with five levels of criticality (A to E). Level A is the most critical (corresponding to the possibility of catastrophic failure) whilst Level E is the least critical (corresponding to a negligible impact on the aircraft and crew).

The relationship between the development of aircraft hardware and software is shown in Figure 13.2. Note that the two life cycles (hardware and software) are closely interrelated simply because a change in hardware configuration inevitably requires a corresponding change to the software configuration. The Safety Assessment Process (SAP) is a parallel activity to that of the System Development Process. It is important to be aware that changes to the system design and configuration will always necessitate a re-appraisal of safety factors.

Testing takes place throughout the development process. Independent tests are usually carried out in order to ensure that the results of tests are valid. Testing generally also involves simulation of out-of-range inputs and abnormal situations such as recovery from power failure (ensuring that a system restart is accomplished without generating dangerous or out-of-range outputs).

Traceability of software is a key component of the DO-178B criteria. Planning documents and evidence of traceability help to ensure that not only are certification requirements met but also that the final code contains all of the required modules and that each module is the most recently updated version. Care is also needed to ensure that none of the final code will be detrimental to the overall operation of the system (for example, seeking data from sensors and transducers that may not be fitted in some configurations of a particular aircraft).

13.3 Software upgrading

When considering software modifications and upgrades it is important to distinguish between executable code (i.e. computer programs) and the data that is used by programs but is not, in itself, executable code. Both of these are commonly

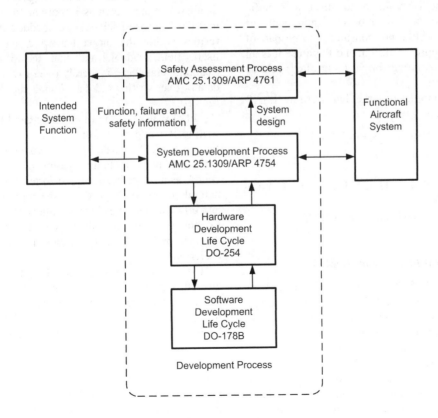

Figure 13.2 The software development process

referred to as 'software' and both are likely to need modification and upgrading during the life of an aircraft.

13.3.1 Field loadable software

Field Loadable Software (FLS) is executable code (i.e. computer programs) that can be loaded into a computer system whilst the system is in place within the aircraft. FLS can be loaded onto an aircraft system by a maintenance mechanic/technician in accordance with defined maintenance manual procedures.

There are three main types of FLS; Loadable Software Aircraft Parts (LSAP), User Modifiable Software (UMS), and Option Selectable Software (OSS).

Loadable software aircraft parts

A Loadable Software Aircraft Part (LSAP) is software that is required to meet a specific airworthiness or operational requirement or regulation. LSAP is not considered to be part of the aircraft approved design and therefore it is an aircraft part requiring formal (controlled) release documentation.

Typical examples of target hardware for LSAP (FLS) include:

- Electronic Engine Controls (EEC)
- Digital Flight Data Acquisition Units (DFDAU)
- Auxiliary Power Unit's Electronic Control Units (ECU)
- Flight Guidance Computers (FGC).

User modifiable software (UMS)

User Modifiable Software (UMS) is declared by the aircraft Type Certificate holder's design organisation (or Supplementary Type Certificate holder's design organisation) as being intended for modification within the constraints established during certification. UMS can usually be upgraded by the aircraft operator, design organisation, or equipment manufacturer, without further review by the licensing authority. Typical examples of target hardware for UMS include:

- Aircraft Condition Monitoring Systems (ACMS)
- In-Flight Entertainment Systems (IFE).

Option selectable software (OSS)

Option Selectable Software (OSS) is software that contains approved and validated components and combinations of components that may be activated or modified by the aircraft operator within boundaries defined by the Type Certificate or Supplementary Type Certificate holder. Typical examples of target hardware for OSS can be found in Integrated Modular Avionics (IMA) units.

13.3.2 Database field loadable data

Database Field Loadable Data (DFLD) is data that is field loadable into target hardware databases. Note that it is important to be aware that the database itself is an embedded item that resides within the target hardware and is not, itself, field loadable and that the process of 'loading a database' is merely one of writing new data or over-writing old data from a supplied data file.

DFLD is usually modified or updated by over-writing it using data from a data file which is field loaded. The data file can contain data in various formats, including natural binary, binary coded decimal, or hexadecimal formats. Of these, natural binary code produces the most compact and efficient data files but the data is not readable by humans and, for this reason, binary coded decimal or hexadecimal formats are sometimes preferred.

It is important to note that the updating of an aircraft database will usually have an impact on aspects of the aircraft's performance. However, if the data used by a program is invalid or has become corrupt, this may result in erratic or out of range conditions. Because of this, it is necessary to treat DFLD in much the same manner as the executable code or LSAP that makes use of it. Hence a DFLD must be given its own unique part number and release documentation.

Typical examples of the target hardware with databases that can be field loaded with DFLD (and that need to be tracked in the same manner as other aircraft parts) include:

- Flight Management Computers (FMC)
- Terrain Awareness Warning System (TAWS) Computers
- Integrated Modular Avionics (IMA) units.

13.3.3 Distribution methods

FLS and DFLD can be distributed by various methods including combinations of the following methods:

- Media distribution. A process whereby FLS or data files are moved from the production organisation or supplier to a remote site using storage media such as floppy disk, a PCMCIA (Personal Computer Memory Card International) card, a CD-ROM, or an Onboard Replaceable Module (OBRM)
- Electronic transfer. A process where a laptop, hand-held computer or Portable Data Loader is used to transfer data using a serial data link or temporary bus connection
- Electronic distribution. A process whereby FLS or DFLD are moved from the producer or supplier to a remote site without the use of intermediate storage media, such as floppy disk or CD-ROM).

Note that the method of release is dependant upon whether the FLS or DFLD is required to meet a specific airworthiness or operational requirement, or certification specification. For FLS or DFLD that does not need to meet a specific airworthiness, operational or certification requirement, a Certificate of Conformity is normally sufficient. In other cases an EASA Form 1 or FAA 8130-3 should accompany any FLS (executable code) that is required to meet a specific airworthiness or operational requirement or regulation, or certification specification, i.e. LSAP. Examples of LSAP that would require such release would include Electronic Engine Controls (EEC), Digital Flight Data Acquisition Units (DFDAU), Auxiliary Power Unit's Electronic Control Units (ECU), Flight Guidance

Computers (FGC), and Integrated Modular Avionics (IMA) units.

An EASA Form 1 or FAA 8130-3 should accompany any DFLD that is required to meet a specific airworthiness or operational requirement or regulation, or certification specification. Examples of DFLD that require such release include Flight Management Computers (FMC), Terrain Awareness Warning System (TAWS) Computers, Integrated Modular Avionics (IMA) units. A 'Letter of Acceptance' or equivalent should accompany the release of any navigational database's DFLD because an EASA Form 1 or FAA 8130-3 cannot be provided.

By virtue of the speed of distribution and the removal of the need for any physical transport media (which can be prone to data corruption), electronic distribution is increasingly being used to transfer FLS or DFLD from the supplier to an operator. Operators should maintain a register which provides the following information:

- The current version of the FLS and DFLD installed
- Which aircraft the FLS and DFLD are installed on
- The aircraft, systems and equipment that they are applicable to
- The functions that the recorded FLS or DFLD performs
- Where (including on or off aircraft location), and in what format it is stored (i.e. storage media type), the name of the person who is responsible for it, and the names of those who may have access to it
- Who can decide whether an upgrade is needed and then authorise that upgrade

---

Key Point

If there is any doubt about the status of an FLS or DFLD upgrade (for example, if there is a reason to suspect that the data may be corrupt or may not be the correct version for the particular aircraft configuration), no attempt should be made to transfer the data. Instead, it should be quarantined and held pending data integrity and/or version checks specified by the supplier.

- A record of all replicated FLS/DFLD, traceable to the original source.

When transferring field loadable software or a database, it is essential to ensure that:

- The FLS or DFLD has come from an appropriate source
- Effective configuration control processes are in place to ensure that only the correct data and/or executable code will be supplied
- The FLS or DFLD is accompanied by suitable release documentation and records are kept.
- Suitable controls are in place to prevent use of FLS and DFLD that have become corrupted during its existence in any 'open' environment, such as on the Internet or due to mishandling in transit
- Effective data validation and verification procedures are in place
- The FLS and DFLD as well as the mechanisms for transferring them (file transfer and file compaction utilities) are checked for unauthorised modification (for example, that caused by malicious software such as viruses and 'spyware').

Special precautions are needed when the FLS or DFLD is transferred or downloaded using the public Internet. These precautions include the use of an efficient firewall to prevent unauthorised access to local servers and storage media as well as accredited virus protection software. Further precautions are necessary when making backup copies of transferred or downloaded data. In particular, backup copies should be clearly labelled with version and date information and they must always be stored in a secure place.

Key Point

When an FLS or DFLD is transferred over an 'open environment' (such as the Internet) extra care will be needed to ensure that the software has not become corrupt or contaminated by viruses. The supplier will usually be able to provide guidance on approved methods of checking software and data integrity.

Test your understanding 13.2

1. Explain what is meant by a Loadable Software Aircraft Part (LSAP).
2. Give TWO examples of an LSAP.

Test your understanding 13.3

State FOUR precautions that must be taken when transferring Field Loadable Software.

Test your understanding 13.4

Explain why Database Field Loadable Data (DFLD) should be treated with the same care and consideration as Field Loadable Software (FLS).

A typical software loading procedure for the Electronic Engine Control (EEC) of a modern passenger aircraft (see Fig. 13.3) is as follows:

1. Make sure the electrical power to the EEC is off
2. Connect one end of the Portable Data Loader (PDL) cable to the connector of the loader (see Fig. 13.4)
3. Disconnect the electrical connector from the EEC
4. Connect the other end of the PDL cable to the connector of the EEC
5. Connect the 28 VDC power source to the Brown (+), Black (−), and shield (chassis ground) terminals of the loader cable
6. Load the software using the following steps:
   (a) Check that you have the correct disk that will be needed to load channels A and B with the non-volatile memory (NVM) and the operating program software
   (b) Turn on the 28 VDC supply to the loader
   (c) Turn on 115 VAC (400 Hz) to the EEC
   (d) Turn on the loader without the disk inserted. The PDL display will show DISK NOT INSERTED after the PDL is initiated
   (e) Make sure you use the correct EEC

software version disk and also ensure that the same software version is used for both the left and right engines

(f) Check that the software disk is not write protected before you load the new software (the disk is write protected as shown in Fig. 13.5)

(g) Insert the appropriate disk into the PDL drive within five minutes after you turned on the power

(h) The PDL will automatically load the NVM and the operating program software to the two channels of the EEC. If it is successful, the PDL will display LOAD COMPLETE. It will take approximately 12 to 16 minutes to load the EEC. If the software loading has failed, it may take a few minutes before the transfer fail lamp illuminates. If the transfer fail lamp illuminates, make sure the cable connections are correct. Remove the power and repeat the software loading steps above. Note that no more than three attempts can be made if the failure reccurs

(i) If LOAD COMPLETE is displayed on the loader, turn off the loader and then turn off the 115 VAC or 28 VDC

(j) Remove the disk from the loader.

7. Remove all electrical power from the EEC and the loader

8. Remove the loader and reconnect the EEC cables

9. Use the appropriate memory verify program (included in the software reprogramming disks) to make sure the EEC has been reprogrammed correctly. If the verify program displays PASS for the EEC part number as well as the checksums, the EEC has been programmed correctly. If FAIL is displayed because the comparison of the FADEC /EEC hardware part number did not agree with the software part number, make sure the FADEC / EEC hardware part number is correct on the nameplate. Use the PRINTSCREEN feature of the PC, and make a paper copy of the final results

10. Use a ball point pen to write the number of the appropriate software version, Service Bulletin number and date on the Service Bulletin plate (a metallic sticker) on the EEC (see Fig. 13.3).

Figure 13.3 Electronic Engine Control (EEC) requiring Field Loadable Software (FLS)

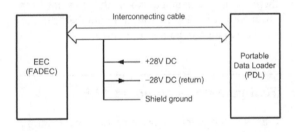

Figure 13.4 Typical Portable Data Loader (PDL) arrangement

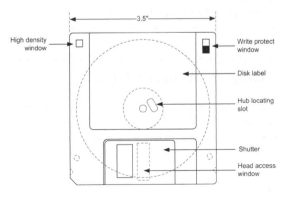

Figure 13.5 Floppy disk with write protect window (the window must be closed to write to the disk)

13.4 Data verification

Various techniques are used to check data files and executable code in order to detect errors. Common methods involve the use of **checksums** and **cyclic redundancy checks** (CRC). Both of these methods will provide an indication that a file has become corrupt, but neither is completely foolproof.

Checksums involve adding the values of consecutive bytes or words in the file and then appending the generated result to the file. At some later time (for example when the file is prepared for loading) the checksum can be re-calculated and compared with the stored result. If any difference is detected the file should not be used.

Cyclic redundancy checks involve dividing consecutive blocks of binary data in the file by a specified number. The remainder of the division is then appended to the file as a series of **check digits** (in much the same way as a checksum). If there is no remainder when the file is later checked by dividing by the same number, the file can be assumed to be free from errors.

Test your understanding 13.5

Classify each of the following applications in terms of level of software criticality:

1. Weather radar
2. VOR
3. In-flight entertainment.

Test your understanding 13.6

An FLS upgrade may have been corrupted during transfer. What action should be taken?

Test your understanding 13.7

Distinguish between User Modifiable Software (UMS) and Option Selectable Software (OSS). Give a typical example of each.

Test your understanding 13.8

(a) Describe TWO methods of checking that a data file contains no errors.

(b) Describe the precautions that should be taken when making backups of FLS.

13.5 Multiple choice questions

1. A Level C software classification is one in which a failure could result in:
   (a) aircraft loss
   (b) fatal injuries to passengers or crew
   (c) minor injuries to passengers or crew.

2. A Level B software classification is one in which the probability of failure is:
   (a) extremely improbable
   (b) extremely remote
   (c) remote.

3. A software Configuration Management Plan (CMP) must be created and maintained by:
   (a) the CAA or FAA
   (b) the aircraft operator
   (c) the relevant DO.

4. Weather radar is an example of:
   (a) Class B software
   (b) Class C software
   (c) Class D software.

5. Electronic Engine Control (EEC) software is an example of:
   (a) DFLD
   (b) LSAP
   (c) OSS.

6. OSS and UMS are specific classes of:
   (a) FLS
   (b) LSAP
   (c) DFLD.

7. The final stage of loading EEC software is:
   (a) disconnecting the PDL
   (b) verifying the loaded software
   (c) switching on and testing the system.

Chapter 14 EMC

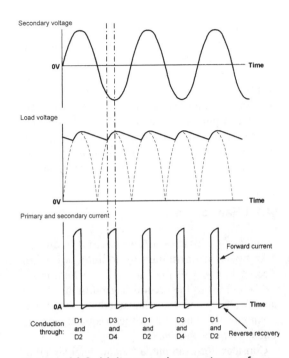

Figure 14.2 Voltage and current waveforms for the circuit shown in Figure 14.1

It is an unfortunate fact of life that the operation of virtually all items of electrical and electronic equipment can potentially disturb the operation of other nearby items of electronic equipment. In recent years increasing incidence of interference has prompted the introduction of legislation that sets strict standards for the design of electrical and electronic equipment. The name given to this type of disturbance is Electromagnetic Interference (EMI) and the property of an electrical or electronic product (in terms of its immunity to the effects of EMI generated by other equipment and its susceptibility to the generation and radiation of its own EMI) is known as Electromagnetic Compatibility (EMC).

EMI (and the need for strict EMC control) can be quickly demonstrated by placing a portable radio receiver close to a computer and tuning over the AM and FM wavebands. Wideband noise and stray EMI radiation should be easily detected.

14.1 EMI generation

As an example of how EMI can be produced from even the most basic of electronic circuits, consider the simple DC power supply consisting of nothing more than a transformer, bridge rectifier, reservoir capacitor and load resistor (see Fig. 14.1). At first sight a simple circuit of this type may look somewhat benign but just take a look at the waveforms shown in Figure 14.2.

The primary and secondary voltage waveforms are both sinusoidal and, as you might expect, the load voltage, comprises a DC level (just less than the peak secondary voltage) onto which is

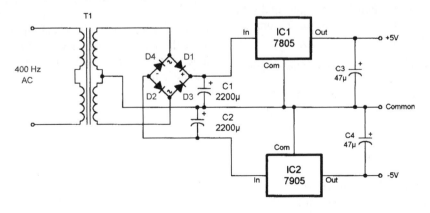

Figure 14.1 Circuit diagram of a simple regulated DC power supply

superimposed a ripple component at 800 Hz. What's more significant (in terms of EMC and EMI) is the waveform of the current that flows in both the secondary and primary circuits. Rather than being sinusoidal (as you might have thought) this current comprises a series of fast rise-time rectangular pulses as each pair of diodes conducts alternately in order to replace the lost charge in the reservoir capacitor. Unfortunately, these rectangular pulses contain numerous harmonics.

14.1.1 Harmonics

An integer multiple of a **fundamental** frequency is known as a **harmonic**. In addition, we often specify the order of the harmonic (second, third, and so on). Thus the **second harmonic** has twice the frequency of the fundamental, the **third harmonic** has three times the frequency of the fundamental, and so on.

Consider, for example, a fundamental signal at 1 kHz. The second harmonic would have a frequency of 2 kHz, the third harmonic a frequency of 3 kHz, and the fourth harmonic a frequency of 4 kHz. Note that, in musical terms, the relationship between notes that are one *octave* apart is simply that the two frequencies have a ratio of 2:1 (in other words, the higher frequency is double the lower frequency).

All complex waveforms (of which rectangular pulses and square waves are examples) comprise a fundamental component together with a number of harmonic components, each having a specific amplitude and with a specific phase relative to the fundamental. The mathematical study of complex waves is known as 'Fourier analysis' and this allows us to describe a complex wave using an equation of the form:

$$v = V_1 \sin(\omega t) + V_2 \sin(2\omega t \pm \varphi_2) + V_3 \sin(3\omega t \pm \varphi_3) + ...$$

where v is the instantaneous voltage of the complex waveform at time, t. V_1 is the amplitude of the fundamental, V_2 is the amplitude of the second harmonic, V_3 is the amplitude of the third harmonic, and so on. Similarly, φ_2 is the phase angle of the second harmonic (relative to the fundamental), φ_3 is the phase angle of the third harmonic (relative to the fundamental), and so on. The important thing to note from this is that all of the individual components that go to make up a complex waveform have a sine wave shape. Putting this another way, a complex wave is made up from a number of sine waves.

14.1.2 Frequency spectrum of a pulse

A rectangular pulse comprises a fundamental component together with an infinite series of harmonics. Taking a square wave as an extreme example of a rectangular pulse, the composite waveform can be analysed into the following components:

- A fundamental component at a frequency, f, and amplitude, V
- A third harmonic component at a frequency, $3f$, and amplitude $V/3$
- A fifth harmonic component at a frequency, $5f$, and amplitude, $V/5$
- A seventh harmonic component at a frequency, $7f$, and amplitude, $V/7$.

and so on, ad-infinitum!

This process (up to the seventh harmonic) is shown in Figure 14.3. The corresponding frequency spectrum (showing amplitude plotted against frequency) appears in Figure 14.5. Note that, to produce a perfect square wave, the amplitude of the harmonics should decay in accordance with their harmonic order and they must all be in phase with the fundamental.

Using Fourier analysis, the equation for a square wave voltage is:

$$v = V \sin(\omega t) + \frac{V}{3}\sin(3\omega t) + \frac{V}{5}V \sin(5\omega t) + ...$$

where V is the amplitude of the fundamental and ω is the angular frequency of the fundamental (note that $\omega = 2\pi f$, where f is the frequency of the fundamental).

Test your understanding 14.1

A square wave voltage has an amplitude of 50 V and a frequency of 2 kHz. Determine:

(a) the amplitude of the fifth harmonic component
(b) the frequency of the fifth harmonic component.

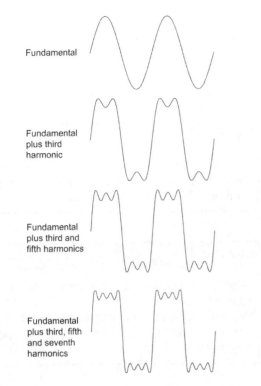

Figure 14.3 Constituents of a square wave

14.2 EMC and avionic equipment

In recent years, EMC and EMI have become a very important consideration for avionic equipment designers. To ensure compliance with increasingly demanding standards, it has become essential for designers to consider the effects of unwanted signals generated by avionic equipment as well as the susceptibility of the equipment to interference from outside. To illustrate this important point we will again use the example of the simple low-voltage DC power supply that we met in Figure 14.1. For this unit to meet stringent EMC requirements it needs to be modified as shown in Figure 14.4. The additional components have the following functions:

1. C5, C6, C7 and C8 provide additional decoupling (effective at high-frequencies) in order to prevent instability in IC1 and IC2. Without these components, and depending upon circuit layout (stray reactances) the regulator circuits may oscillate at a high-frequency.
2. C9 and C10 provide additional high-frequency decoupling to remove noise present on the output voltage rails.
3. C11, L1, L2, C12 and C13 provide a low-pass supply filter to remove noise and spurious signals resulting from the harmonics of the switching action of the diode rectifiers. This filter also reduces supply borne noise that would otherwise enter the equipment from the supply.
4. A low-resistance ground connection is introduced to ensure that there is an effective connection between aircraft ground and the equipment chassis (note that there is also an earth connection to the laminations and internal screening on the mains transformer).

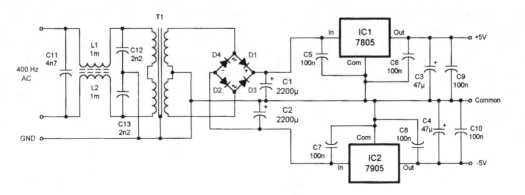

Figure 14.4 The modified DC power supply with significantly improved EMC performance

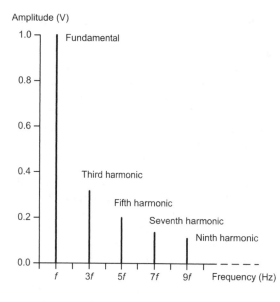

Figure 14.5   The frequency spectrum of a square wave

14.3 Spectrum analysis

Accurate EMC measurements involve the use of a specialised instrument known as a **spectrum analyser**. Such instruments display signal level (amplitude) against frequency. The complete block diagram of a modern spectrum analyser is shown in Figure 14.6. The radio frequency (RF) input signal is fed to a relay switched preamp/attenuator section that provides a maximum of 50 dB attenuation or up to 10 dB of gain. The signal then passes through a low-pass filter, into a wideband mixer where it is mixed with the 1.5 to 2.5 GHz local oscillator signal to give an intermediate frequency (IF) output at 1.5 GHz.

The tuning/sweep voltage is fed through a linearising amplifier which compensates for the oscillator's non linear tuning characteristic. The tuning input is 0 V at zero input frequency, up to +5 V at 1000 MHz input frequency. This represents a tuning characteristic equivalent to 200 MHz/V.

The signal from the first mixer is then amplified by the first IF amplifier and then filtered by a four-stage band pass 1500 MHz cavity filter. The output of the filter is then coupled into the second mixer where it is mixed

with the output of the second local oscillator to give the second IF output at 10.7 MHz. The second local oscillator is tuned to 10.7 MHz below the first IF (i.e. approx. 1490 MHz).

Next, the 10.7 MHz IF signal is fed to a wideband amplifier. In 'wideband mode' the IF signal is passed to a logarithmic amplifier. The output of the wideband amplifier also feeds the 120 kHz filter/amplifier stages. The combined affect of a total of three ceramic filters provides the required filter shape. The output of the 120 kHz filter stage also feeds the 9 kHz filter stages.

The logarithmic amplifier consists of an initial stage followed by two 30 dB gain limiting amplifiers. Audio signal monitoring is provided by means of a conventional audio amplifier. The input of this stage can be taken from the FM detector or from the logarithmic amplifier when AM demodulation is required.

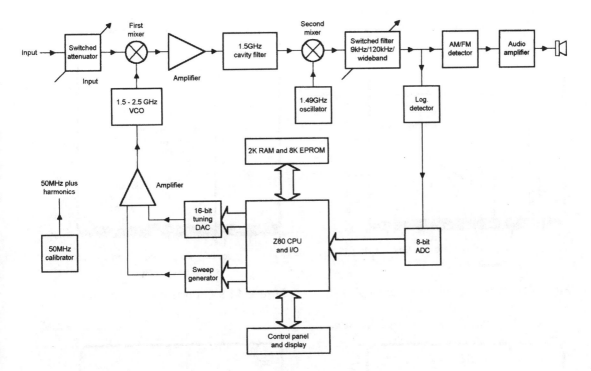

Figure 14.6 Block diagram of a spectrum analyser

An 8-bit Z80 microprocessor carries out the main digital signal processing and also generates the digital video signals that are fed to the display circuitry. The digital system's master clock operates at 8 MHz which is then divided by two to provide the Z80's clock input. Further division of the master clock provides the display system's line frequency (31.25 kHz) as well as the frame signal (60 Hz) and a further 30 Hz control signal.

The Z80 CPU operates in conjunction with an 8K EPROM containing the control software and a 2K RAM that stores the video and calibration data. In order to retain the calibration data when the unit is switched off, the RAM is backed up by a long-life lithium battery.

The video display system uses a vertical raster with 448 lines displayed, each having a vertical resolution of 192 pixels (bits). The line scan is from bottom to top, the frame scan from left to right. The picture is made up of the spectrum display together with separately enabled amplitude and frequency cursors.

An internal 50 MHz crystal calibrator (using a third overtone crystal) is used to provide calibration at harmonics up to 1 GHz. The fundamental output of this stage is set accurately to a level of −20 dBm.

A typical display produced by the analyser is shown in Figure 14.7. The vertical scale shows amplitude in dBm (decibels relative to 1 mW) over the range −100 dBm to +10 dBm, whilst the horizontal scale shown frequency (in MHz) over the range 0.1 MHz to 1,000 MHz. The signal under investigation appears as a series of lines (i.e. a line spectrum). In order to provide a means of accurately measuring the amplitude and frequency of individual signal components, the spectrum analyser display has two cursors; one for amplitude and one for frequency. The measured signal component in Figure 14.7 has an amplitude of −26 dBm and a frequency of 121 MHz (note that modern instruments invariably provide separate digital displays of the cursor settings).

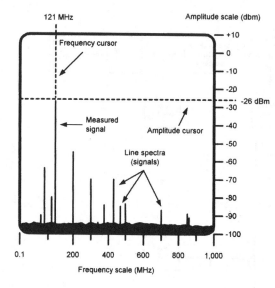

Figure 14.7   Spectrum analyser display

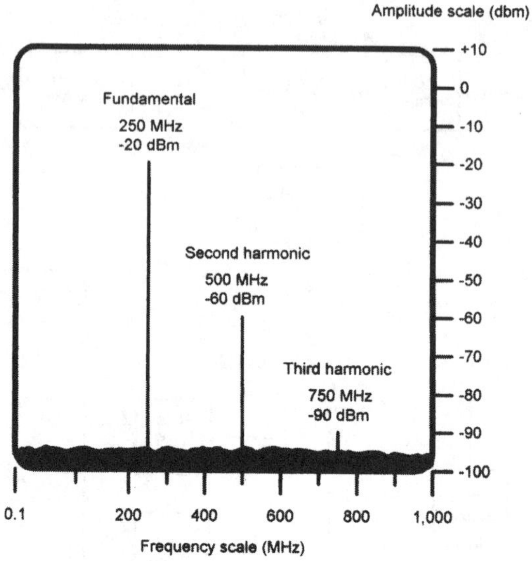

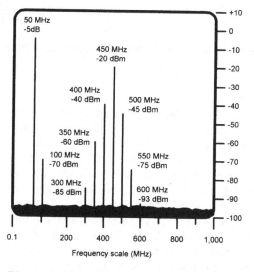

Figure 14.9   Fundamental and harmonics

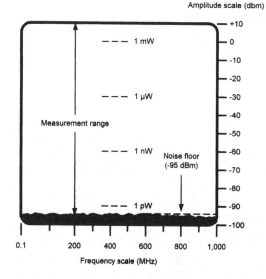

Figure 14.8   Measurement range

Figure 14.10   Modulated signal components

The measurement range provided by the spectrum analyser amounts to a total of 110 dB but the ultimate sensitivity of the instrument is determined by the noise produced in the instrument itself. This is usually referred to as instrument's **noise floor** (see Fig. 14.8). In this case, the noise floor (with no signal connected) is at a level of about −95 dB (less than 1 picowatt).

Note that modern instruments typically have noise floors of between −95 and −110 dB.

Figure 14.9 shows how the spectrum analyser displays a fairly simple complex waveform comprising a fundamental component at 250 MHz with a level of −20 dBm, a second harmonic at 500 MHz with a level of −60 dBm, and a third harmonic at 750 MHz with a level of −90 dBm.

Note that any other harmonic components that may be present have a level that is below the instrument's noise floor (i.e. they must have signal levels of less than −95 dBm).

The display of a more complex signal is shown in Figure 14.10. Here there are two signals—one modulated by the other—together with a number of side frequency components and harmonics. The waveform has the following components:

- a fundamental component at 50 MHz with a level of −5 dBm
- a higher frequency fundamental at 450 MHz with a level of −20dB modulated by the 50 MHz signal
- a first pair of side frequency components at 400 MHz and 500 MHz with levels of −40 dBm and −45 dBm respectively
- a second pair of side frequency components at 350 MHz and 550 MHz with levels of −60 dBm and −75 dBm respectively
- a third pair of side frequency components at 300 MHz and 600 MHz with levels of −85 dBm and −93 dBm respectively
- a second harmonic of the 50 MHz fundamental at 100 MHz with a level of −70 dBm.

A signal with noise sidebands is shown in Figure 14.11. Since noise is essentially a randomly distributed waveform (in terms of frequency) it appears as a blurred area (like the noise floor itself) rather than as a line spectrum. This diagram shows how the noise spreads out either side of the fundamental at 450 MHz. The second harmonic of the fundamental (at 900 MHz) is also just discernible above the noise.

Figures 14.12 and 14.13 respectively show the appearance of low and high-frequency noise. In the first case, the noise reaches a broad peak at about 10 MHz whilst in the latter it reaches a peak at about 900 MHz.

14.3.1 Frequency bands

For convenience the complete frequency spectrum is divided into a number of bands. These bands are often referred to when describing the effects of particular types of EMI and they are shown in Table 14.1.

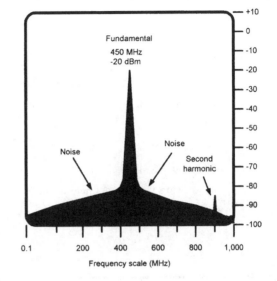

Figure 14.11 Noise sidebands

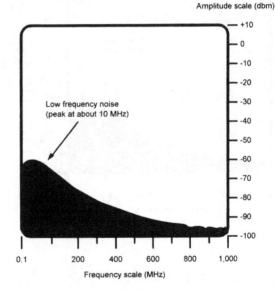

Figure 14.12 Low frequency noise

Test your understanding 14.3

Explain what is meant by the 'noise floor' of a spectrum analyser and the effect that this will have on the ability of the instrument to accurately measure EMI.

Table 14.1 Frequency bands used in EMI measurement

Frequency range	Wavelength	Designation
300 Hz to 3 kHz	1,000 km to 100 km	Extremely low frequency (ELF)
3 kHz to 30 kHz	100 km to 10 km	Very low frequency (VLF)
30 kHz to 300 kHz	10 km to 1 km	Low frequency (LF)
300 kHz to 3 MHz	1 km to 100 m	Medium frequency (MF)
3 MHz to 30 MHz	100 m to 10 m	High frequency (HF)
30 MHz to 300 MHz	10 m to 1 m	Very high frequency (VHF)
300 MHz to 3 GHz	1 m to 10 cm	Ultra high frequency (UHF)
3 GHz to 30 GHz	10 cm to 1 cm	Super high frequency (SHF)

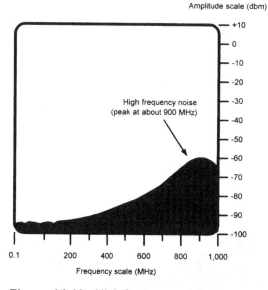

Figure 14.13   High frequency noise

14.4 Effects and causes of EMI

Electromagnetic Interference (EMI) can be defined as the presence of unwanted voltages or currents which can adversely affect the performance of an avionic system. The effects of EMI include errors in instrument indications (both above and below true), heterodyne whistles present on audio signals, herringbone patterns in video displays, repetitive pulse noise (buzz) on intercom and cabin phone systems, desensitising of radio and radar receivers, false indications in radar and distance measuring equipment, unwarranted triggering of alarms, and so on. Note that, in certain circumstances, the performance of the device that emits EMI may also suffer impaired performance.

14.4.1 Sources of EMI

Some sources known to emit EMI include fluorescent lights, radio and radar transmitters, power lines, window heat controllers, induction motors, switching and light dimming circuits, microprocessors and associated circuitry, pulsed high frequency circuits, bus cables, (but *not* fibre optic cables), static discharge and lightning. The energy generated by these sources can be conducted and/or radiated as an electromagnetic field. Unless adequate precautions are taken to eliminate the interference at source and/or to reduce the equipment's susceptibility to EMI, the energy can then become coupled into other circuits.

Conduction is the process in which the energy is transmitted through electrically conductive paths such as circuit wiring or aircraft metallic structure. In electromagnetic field **radiation** energy is transmitted through electrically non-conductive paths, such as air, plastic materials, or fibreglass.

Systems which may be susceptible to electromagnetic interference include radio and radar receivers, microprocessors and other microelectronic systems, electronic instruments, control systems, audio and in-flight entertainment systems (IFE).

Whether a system will have an adverse response to electromagnetic interference depends on the type and amount of emitted energy in conjunction with the susceptibility threshold of the receiving system. The threshold of susceptibility is the minimum interference signal level (conducted or radiated) that results in equipment performance that is indistinguishable from the normal response. If the threshold is exceeded then the performance of the equipment will become degraded. Note that, when the susceptibility threshold level is greater than the levels of conducted or radiated emissions, electromagnetic interference problems do not exist. Systems to which this applies are said to be Electromagnetically Compatible. In other words, the systems will operate as intended and any EMI generated is at such a level that it does affect normal operation.

14.4.2 Types of interference

EMI can be categorised by bandwidth, amplitude, waveform and occurrence. The bandwidth of interference is the frequency range in which the interference exists. The interference bandwidth can be narrow or broad.

Narrow band interference can be caused by such items as AC power rails, microprocessor clocks (and their harmonics), radio transmitters and receivers. These items of equipment all contain sources (e.g. clock oscillators) that work on specific frequencies. These signals (along with unwanted harmonics) can be radiated at low levels from the equipment.

Broad band interference is caused by devices generating random frequencies and noise which may be repetitive but is not confined to a single frequency or range of frequencies. Examples of this type of interference are power supplies, LCD and AMLCD (by virtue of their high-frequency AC supplies), switched mode power supplies, switching power controllers, and microprocessor bus systems.

Interference amplitude is the strength of the

signal received by the susceptible system. The amplitude can be constant or can vary predictably with time, or can be totally random. For example, a 115 V AC power line can induce a stable sinusoidal waveform on adjacent 28 V DC power or signal lines. The amplitude of the interference will depend on the load current in the AC power line (recall that the magnetic field produced around a conductor is directly proportional to the current flowing in the conductor). Examples of random interference are environmental noise and inductive switching transients. Environmental noise is the aggregate of all electromagnetic emissions present in a particular space or area of concern at any one time. This is usually measured over a defined spectrum (e.g. 30 kHz to 30 MHz).

It is important to be aware that there is no one specific waveform that produces electromagnetic interference. Instead, it is the change from one signal level to another in conjunction with the rate at which it changes that determines the amount of electromagnetic energy released. More energy is released when the change in signal level and rate is increased.

The occurrence of EMI can be categorized as being either **periodic** (continuously repetitive), **aperiodic** (predictable but not continuous), or **random** (totally unpredictable).

14.4.3 EMI reduction

Planning for electromagnetic compatibility must be initiated in the design phase of a device or system (as discussed in 14.2). If this is not satisfactorily achieved, interference problems may arise. The three factors necessary to produce an EMI problem are a noise source (see Fig. 14.14), a means of coupling (by conduction or radiation), and a susceptible receiver. To reduce the effects of EMI, *at least one* of these factors must be addressed. The following lists techniques for EMI reduction under these three headings (note that some techniques address more than one factor):

1. Suppress interference at source

- Enclose interference source in a screened metal enclosure and then ensure that the enclosure is adequately grounded

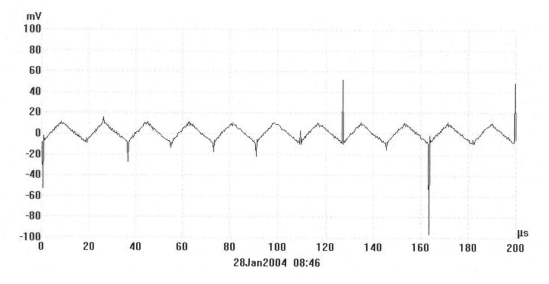

Figure 14.14 Transient switching pulses appearing on the output of a DC power supply without adequate supply filtering. Note that the AC ripple is 20 mV peak-peak whilst the noise spikes are up to five times this value. The noise spikes can be virtually eliminated by adding a supply filter.

- Use transient suppression on relays, switches and contactors
- Twist and/or shield bus wires and data bus connections
- Use screened (i.e. coaxial) cables for audio and radio frequency signals
- Keep pulse rise times as slow and long as possible
- Check that enclosures, racks and other supporting structures are grounded effectively.

2. Reduce noise coupling

- Separate power leads from interconnecting signal wires.
- Twist and/or shield noisy wires and data bus connections
- Fit an optical fibre data bus where possible
- Use screened (i.e. coaxial) cables for audio and radio frequency signals
- Keep ground leads as short as possible.
- Break interference ground loops by incorporating isolation transformers, differential amplifiers, balanced circuits.
- Filter noisy output leads.
- Physically relocate receivers and sensitive equipment away from interference source.

3. Increasing susceptibility thresholds

- Limit bandwidth to only that which is strictly necessary
- Limit gain and sensitivity to only that which is strictly necessary
- Ensure that enclosures are grounded and that internal screens are fitted
- Fit components that are inherently less susceptible to the effect of stray radiated fields.

14.5 Aircraft wiring and cabling

When many potential sources of EMI are present in a confined space, aircraft wiring and cabling has a crucial role to play in maintaining electromagnetic compatibility. The following points should be observed:

1. Adequate wire separation should be maintained between noise source wiring and susceptible wiring (for example, ADF wiring should be strategically routed in the aircraft to ensure a high level of EMC).

2. Any changes to the routing of this wiring could have an adverse affect on the system. In

addition, the wire separation requirements for all wire categories must be maintained.

3. Wire lengths should be kept as short as possible to maintain coupling at a minimum. Where wire shielding is incorporated for lightning protection, it is important that the shield grounds (pigtails) be kept to their designed length. An inch or two added to the length will result in degraded lightning protection.

4. Equipment grounds must not be lengthened beyond design specification. A circuit ground with too much impedance may no longer be a true ground.

5. With the aid of the technical manuals, grounding and bonding integrity must be maintained. This includes proper preparation of the surfaces where electrical bonding is made.

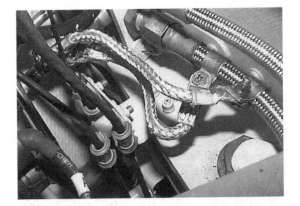

Figure 14.15 Bonding straps

14.6 Grounding and bonding

The electrical integrity of the aircraft structure is extremely important as a means of reducing EMI and also protecting the aircraft, its passengers, crew and systems, from the effects of lightning strikes and static discharge. Grounding and bonding are specific techniques that are used to achieve this (see Fig. 14.15). Grounding and bonding can also be instrumental in minimising the effects of high intensity radio frequency fields (HIRF) emanating from high power radio transmitters and radar equipment. Grounding and bonding resistances of less than 0.001 Ω to 0.003 Ω are usually required.

14.6.1 Grounding

Grounding is defined as the process of electrically connecting conductive objects to either a conductive structure or some other conductive return path for the purpose of safely completing either a normal or fault circuit. Bonding and grounding connections are made in an aircraft in order to accomplish the following:

1. Protect aircraft, crew and passengers against the effects of lightning discharge
2. Provide return paths for current

3. Prevent the development of RF voltages and currents
4. Protect personnel from shock hazards
5. Maintain an effective radio transmission and reception capability
6. Prevent accumulation of static charge.

The following general procedures and precautions apply when making bonding or grounding connections:

1. Bond or ground parts to the primary aircraft structure where possible
2. Make bonding or grounding connections so that no part of the aircraft structure is weakened
3. Bond parts individually if feasible
4. Install bonding or grounding connections against smooth, clean surfaces
5. Install bonding or grounding connections so that vibration, expansion or contraction, or relative movement in normal service will not break or loosen the connection
6. Check the integrity and effectiveness of a bonded or grounded connection using an approved bonding tester.

14.6.2 Bonding

Bonding refers to the electrical connecting of two or more conducting objects that are not otherwise adequately connected. The main types of bonding are:

1. **Equipment bonding**. Low impedance paths

to the aircraft structure are generally required for electronic equipment to provide radio frequency return circuits and to facilitate reduction in EMI.

2. **Metallic surface bonding**. All conducting objects located on the exterior of the airframe should be electrically connected to the airframe through mechanical joints, conductive hinges, or bond straps, which are capable of conducting static charges and lightning strikes.

3. **Static bonds**. All isolated conducting paths inside and outside the aircraft with an area greater than 3 in² and a linear dimension over 3 inches that are subjected to electrostatic charging should have a mechanically secure electrical connection to the aircraft structure of adequate conductivity to dissipate possible static charges.

Key Point

Initial control of EMI is achieved in modern aircraft by careful design and rigorous testing. Routine maintenance helps to ensure that the aircraft retains electromagnetic compatibility, thereby keeping EMI problems to a minimum.

Key Point

Effective grounding and bonding provide a means of ensuring the electrical integrity of the aircraft structure as well as minimising the effects of HIRF fields and the hazards associated with lightning and static discharge.

14.7 Multiple choice questions

1. Aperiodic noise is:
   (a) regular but not continuous
   (b) regular and continuous
   (c) entirely random in nature.

2. EMI can be conveyed from a source to a receiver by:
   (a) conduction only
   (b) radiation only
   (c) conduction and radiation.

3. The display produced by a spectrum analyser shows:
   (a) frequency plotted against time
   (b) amplitude plotted against time
   (c) amplitude plotted against frequency.

4. EMI can be reduced by means of:
   (a) screening only
   (b) screening and filtering
   (c) screening, bonding and filtering.

5. The effects of HIRF can be reduced by:
   (a) screening only
   (b) screening and filtering
   (c) screening, bonding and filtering.

6. The typical maximum value of bonding resistance is:
   (a) less than $0.005 \ \Omega$
   (b) between $0.005 \ \Omega$ and $0.05 \ \Omega$
   (c) more than $0.05 \ \Omega$.

7. Effective protection against lightning and static discharge damage to an aircraft requires that:
   (a) all isolated conducting parts must have high resistance to ground
   (b) all parts of the metal structure of the aircraft must be bonded to ground
   (c) all power and bus cables must be well insulated.

8. Supply borne noise can be eliminated by means of:
   (a) a low-pass filter
   (b) a high-pass filter
   (c) a band-pass filter.

9. Noise generated by a switching circuit is worse when:
   (a) switching is fast and current is low
   (b) switching is slow and current is high
   (c) switching is fast and current is high.

Avionic systems

This final chapter provides an overview of the major avionic systems fitted to a modern passenger aircraft. The aim is to put into context the material contained in previous chapters which underpins the operation of these complex systems including:

- Aircraft Communication Addressing and Reporting System (ACARS)
- Electronic Centralised Aircraft Monitoring (ECAM)
- Electronic Flight Instrument System (EFIS)
- Engine Indication and Crew Alerting System (EICAS)
- Fly by Wire (FBW)
- Flight Management System (FMS)
- Global Positioning System (GPS)
- Inertial Reference System (IRS)
- Inertial Navigation System (INS)
- Traffic Alert Collision Avoidance System (TCAS).

This chapter also introduces the test techniques used to detect faults on these systems:

- Automatic Test Equipment (ATE)
- Built-in Test Equipment (BITE).

15.1 ACARS

ACARS (Aircraft Communication Addressing and Reporting System) is a digital data link system transmitted in the VHF range (129 MHz to 137 MHz). ACARS provides a means by which aircraft operators can exchange data with an aircraft without human intervention. This makes it possible for an airline to communicate with the aircraft in their fleet in much the same way as it is possible to exchange data using a land-based digital network. ACARS uses an aircraft's unique identifier and the system has some features that are similar to those currently used for electronic mail.

The ACARS system was originally specified in the ARINC 597 standard but has been revised as ARINC 724B. A significant feature of ACARS is the ability to provide real-time data on the ground relating to aircraft performance (see Fig. 15.1). This has made it possible to identify and plan aircraft maintenance activities.

ACARS communications are automatically directed through a series of ground based ARINC (Aeronautical Radio Inc.) computers to the relevant aircraft operator. The system helps to reduce the need for mundane HF and VHF voice messages and provides a system which can be logged and tracked. Typical ACARS messages cater for the transfer of routine information such as:

- passenger loads
- departure reports
- arrival reports
- fuel data
- engine performance data.

This information can be requested by the company and retrieved from the aircraft at periodic intervals or on demand. Prior to ACARS this type of information would have been transferred via VHF voice communication.

ACARS uses a variety of hardware and software components including those which are installed on the ground and those which are present in the aircraft. The aircraft ACARS

```
ACARS mode: E Aircraft reg: N27015
Message label: H1 Block id: 3
Msg no: C36C
Flight id: CO0004
Message content:-
#CFBBY ATTITUDE INDICATOR
MSG 2820121 A 0051 06SEP06 CL H PL
DB FUEL QUANTITY PROCESSOR UNIT
MSG 3180141 A 0024 06SEP06 TA I 23
PL
DB DISPLAYS-2 IN LEFT AIMS
MSG 2394201 A 0005 06SEP06 ES H PL
MSG 2717018
```

Figure 15.1 Example of a downlink ACARS message sent from a Boeing 777 aircraft

components include a **Management Unit** which deals with the reception and transmission of messages via the VHF radio transceiver, and the **Control Unit** which provides the crew interface and consists of a display screen and printer. The ACARS **Ground Network** comprises the ARINC ACARS remote transmitting/receiving stations and a network of computers and switching systems. The ACARS **Command, Control and Management Subsystem** consists of the ground based airline operations and associated functions including operations control, maintenance and crew scheduling.

There are two types of ACARS messages; **downlink** messages that originate from the aircraft and **uplink** messages that originate from ground stations. The data rate is low and messages comprise plain alphanumeric characters. Frequencies used for the transmission and reception of ACARS messages are in the band extending from 129 MHz to 137 MHz (VHF) as shown in Table 15.1. Note that different channels are used in different parts of the world. A typical ACARS message (see Fig. 15.2) consists of:

- mode identifier (e.g. 2)
- aircraft identifier (e.g. G-DBCC)
- message label (e.g. 5U)
- block identifier (e.g. 4)
- message number (e.g. M55A)
- flight number (e.g. BD01NZ)
- message content (see Fig. 15.2).

Figure 15.1 ACARS channels

Frequency	ACARS service
129.125 MHz	USA and Canada (additional)
130.025 MHz	USA and Canada (secondary)
130.450 MHz	USA and Canada (additional)
131.125 MHz	USA (additional)
131.475 MHz	Japan (primary)
131.525 MHz	Europe (secondary)
131.550 MHz	USA, Canada, Australia (primary)
131.725 MHz	Europe (primary)
136.900 MHz	Europe (additional)

```
ACARS mode: 2
Aircraft reg: G-DBCC
Message label: 5U
Block id: 4
Msg no: M55A
Flight id: BD01NZ
Message content:-
01 WXRQ 01NZ/05 EGLL/EBBR .G-DBCC
/TYP 4/STA EBBR/STA EBOS/STA EBCI
```

Figure 15.2 Example of an ACARS message (see text)

15.2 EFIS

As mentioned in Chapter 1, most modern passenger aircraft use **Electronic Flight Instrument System** (EFIS) displays in what has become known as the 'glass cockpit' (see Figure 15.3). EFIS provides large, clear, high-resolution displays which are easy to view under wide variations of ambient light intensity. Displays can be independently selected and configured as required by the captain and first officer and the ability to display information from various data sources in a single display makes it possible for the crew to rapidly assimilate the information that they need. A notable disadvantage of EFIS is a significant increase in EMI (see page 152 and 153). The two most commonly featured EFIS instruments are the **Electronic Horizontal Situation Indicator** (EHSI) and the **Electronic Attitude Direction Indicator** (EADI) as shown in Figures 15.5 and 15.6.

The EFIS uses input data from several sources including:

- VOR/ILS/MLS
- TACAN (see later)
- pitch, roll, heading rate and acceleration data from an Attitude Heading System Reference (AHRS) or conventional vertical gyro
- compass system
- radar altimeter
- air data system
- Distance Measuring Equipment (DME)
- Area Navigation System (RNAV)
- Vertical Navigation System (VNAV)

- Weather Radar System (WXR)
- Automatic Direction Finder (ADF).

A typical EFIS system comprises:

- Primary Flight Display (PFD)
- Navigation Display (ND)
- Display Select Panel
- Display Processor Unit
- Weather Radar Panel
- Multifunction Display
- Multifunction Processor Unit.

15.2.1 EFIS displays

The **Primary Flight Display** (PFD) is a multicolour CRT or AMLCD displaying aircraft attitude and flight control system steering commands including VOR, localizer, TACAN, or RNAV deviation; and glide slope or pre-selected altitude deviation. The PFD provides flight control system mode annunciation, auto-pilot engage annunciation, attitude source annunciation, marker beacon annunciation, radar altitude, decision height set and annunciation, fast-slow deviation or angle-altitude alert, and excessive ILS deviation.

The **Navigation Display** (ND) provides a plan view of the aircraft's horizontal navigation situation. Data includes heading and track, VOR, ILS, or RNAV course and deviation (including annunciation or deviation type), navigation source annunciation, to/from, distance/direction, time-to-go, elapsed time, course information and source annunciation from a second navigation source, and weather radar data. The ND can also be operated in an approach format or an en-route format with or without weather radar information included in the display.

Operational display parameters (such as ground speed, time-to-go, time, and wind direction/speed) can be selected by means of the **Display Select Panel** (DSP).

The **Multifunction Display** (MFD) is a colour CRT or AMLCD (see Chapter 11). Standard functions displayed by the unit include weather radar (in which the colours green, yellow and red are used to indicate increasing levels of storm activity), pictorial navigation map, check lists and other operating data. In the event of a failure of

Figure 15.3 'Glass cockpit' (EFIS) displays in an A320

Figure 15.4 A320 PFD and ND brightness and transfer controls

any of these displays, the required information can be shown on the displays that remain functional.

The **Display Processor** (DPU)/**Multifunction Processor Unit** (MPU) provides sensor input processing and switching for the necessary deflection and video signals, and power for the electronic flight displays. The DPU can drive up

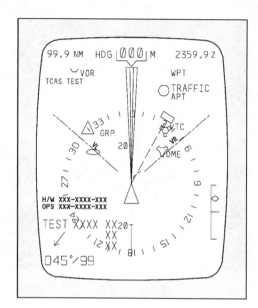

Figure 15.5 Boeing EFIS EHSI display

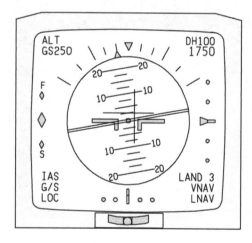

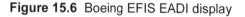

Figure 15.6 Boeing EFIS EADI display

Key Point

ACARS provides the aircraft operator with a means of delivering real-time information on aircraft's performance. The system operates in conjunction with the aircraft's VHF radio and reduces the need for mundane voice traffic.

```
ACARS mode: R Aircraft reg: G-EUPR
Message label: 10 Block id: 8
Msg no: M06A
Flight id: BA018Z
Message content:-
FTX01.ABZKOBA
BA1304
WE NEED ENGINEERING
TO DO PDC ON NUMBER 2
IDG
CHEERS ETL 0740
GMT
```

Figure 15.7 See Test your knowledge 15.1

Test your understanding 15.1

A message from the crew of an A319 aircraft is shown in Figure 15.7. Explain how this message was sent and why it was necessary. Also specify likely frequencies used to transmit the message.

Test your understanding 15.2

Describe two different types of display provided by the EFIS fitted to a modern passenger aircraft. What information appears in each of these displays?

to two electronic flight displays with different deflection and video, signals.

15.2.2 EFIS operation

The simplified block schematic diagram of an EFIS system is shown in Figure 15.8. The system is based on three **symbol generators**. These are simply three microprocessor-based computers (see Chapter 6) that process data received from a wide variety of aircraft systems (VOR, DME, IRS, etc) and generate the signals that drive the EFIS displays (and so generate the 'symbols' that appear on the EFIS screens). The left symbol generator supplies the Captain's EADI and EHSI whilst the right symbol generator supplies the

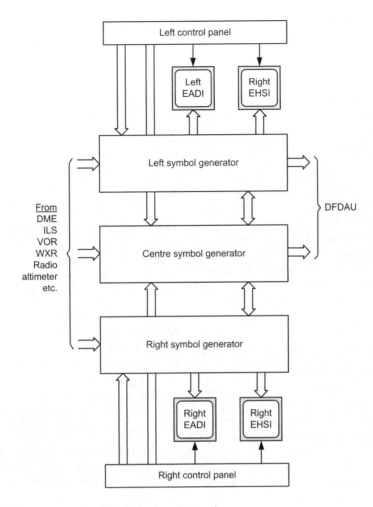

Figure 15.8 EFIS system simplified block schematic

First Officer's EHSI and EADI. The centre symbol generator provides a back-up system that can be used to replace the left or right symbol generator in the event of failure. To cater for the very wide variations in ambient light intensity found in the cockpit, light sensors are fitted to individual displays in order to provide automatic brightness control. Several sensors are usually fitted. For example, two on the glare shield and one on each EADI/EHSI display. In addition, a control panel is fitted in order to provide manual adjustment of the displays by the crew. Outputs from the left and centre symbol generators are sent to the **Digital Flight Data Acquisition Unit** (DFDAU). This unit conditions the signals to

produce a digital output which is sent to the **Digital Flight Data Recorder** (DFDR). The recorded data is stored in a crash protected container inside the DFDR.

The DFDR is turned on automatically when the aircraft is in flight, or on the ground when an engine is running. The system may also be turned on manually by the function switch on the **Data Management Entry Panel** (DMEP) or tested by means of a switch on the flight recorder control panel. The flight recorder system is monitored continuously by **Built-In Test Equipment** (BITE) during operation. System failures are displayed on the flight recorder control panel and on the DFDAU and EICAS.

Each display unit uses either a colour AMLCD or a colour CRT. In the case of a CRT display, two types of scanning are used, **raster scanning** (see page 121) and **stroke scanning**. Raster scanning usually takes place at a rate of 50 Hz or 60 Hz. Stroke scanning is the technique used to display symbols and characters (superimposed on the raster scan). The colour CRT displays use **electromagnetic deflection** and a final anode voltage of around 18 kV. The deflection coils are mounted in a yoke around the neck of the CRT. Note that, by virtue of the resistance of the deflection coils, the scanning voltage waveform has to be trapezoidal in shape rather than the linear ramp waveform (or 'sawtooth' wave) that would be appropriate for electrostatic deflection.

Symbols (see Figure 15.5) are displayed in various colours (for example, white for status symbols, yellow for cautionary information and red for warnings).

15.3 EICAS

The Engine Indication and Crew Alerting System (EICAS) is designed to provide all engine instrumentation and crew annunciations in an integrated format. The equivalent system on Airbus aircraft is the Electronic Centralized Aircraft Monitoring (ECAM) system.

The information supplied by EICAS/ECAM includes display of engine torque, inter-stage turbine temperature, high and low pressure gas generator RPM, fuel flow, oil temperature and pressure. As part of the EICAS, graphical depiction of aircraft systems can be displayed. Such systems as electrical, hydraulic, de-icing, environmental and control surface position can be represented. All aircraft messages relating to these systems are also displayed on the EICAS.

EICAS improves reliability through elimination of traditional engine gauges and simplifies the flight deck through fewer stand-alone indicators. EICAS also reduces crew workload by employing a graphical presentation that can be rapidly assimilated. EICAS can also help to reduce operating costs by providing maintenance data.

A typical EICAS comprises two large high resolution, colour displays together with associated control panels, two or three EICAS data concentrator units and a lamp driver unit.

The primary EICAS display presents primary engine indication instruments and relevant crew alerts. It has a fixed format providing engine data including:

- RPM and temperature
- fuel flow and quantity
- engine vibration
- gear and flap details (where appropriate)
- Caution Alerting System (CAS) messages (colour coded to indicate importance).

The secondary EICAS display indicates a wide variety of options to the crew and serves as a backup to the primary display. They are selectable in pages using the EICAS control panel and include the display of information relating to:

- landing gear position
- flaps/trim
- auxiliary power unit
- cabin pressurisation/anti-ice
- fuel/hydraulics
- flight control positions
- doors/pressurisation/environmental
- AC and DC electrical data.

The EICAS displays receive data bus inputs from the EICAS Data Conversion Unit (DCU). The EICAS displays provide data bus outputs to the Integrated Avionics Processing System (IAPS) Data Concentrator Units (DCU). Note that the pilot or co-pilot can select either display. Selecting one display blanks the second display and allows data pages to be selected. The EICAS control (ECU) panel is used to select pages. The information on the data buses is routed to both EICAS displays and both multifunction displays.

The DCU receives data in various formats from a variety of sensors, including the high and low-speed ARINC 429 bus, from analogue and discrete inputs from the engines and other aircraft systems. These are concentrated and processed for transmission on system buses (ARINC or otherwise).

Outputs include crew alerting logic, engine data to the displays, maintenance, diagnostic and aircraft data to the IAPS DCU, indicator lamp data to the LDU, aircraft system data to the Flight Data Recorder (FDR) and data link management

Figure 15.9 Boeing 757 EICAS

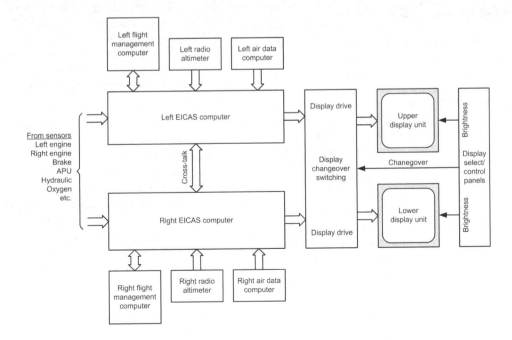

Figure 15.10 EICAS simplified block schematic

unit. The Lamp Driver Unit (LDU) is a dual-channel unit capable of driving up to 120 indicator lamps. Channels 1 and 2 receive digital buses from all the DCU. The buses convey lamp activation words from the DCU. Channels 1 and 2 are identical and the outputs from each side are tied together (wired-OR logic). If one channel lamp sink fails, the other channel lamp sink will provide the function. The LDU monitors the lamp sinks to verify correct function and outputs the lamp sink states on a digital bus to the DCU.

The Electronic Routing Units (ERU) are junction boxes for the data concentrator units. The ERU splits each input signal to three output pins. The pilot ERU routes left-side airplane data and the co-pilot ERU routes right-side airplane data.

EICAS simplifies flight deck clutter by integrating the many electro-mechanical instruments that previously monitored engine and aircraft systems. Safety is increased whilst the pilot workload is simplified. EICAS continuously monitors the aircraft for out-of-tolerance or abnormal conditions and notifies the crew when an event occurs.

15.4 Fly-by-wire (FBW)

Fly-by-wire is the name given to the electrical/electronic flight control system now used in all modern passenger aircraft. FBW was first introduced on a commercial passenger aircraft (the Airbus A320) in 1988.

Rather than mechanical linkages operating hydraulic actuators, fly-by-wire systems move flight control surfaces (ailerons, rudders etc) using electrical wire connections driving motors. At the heart of the system are computers that convert the pilot's commands into electrical signals which are transmitted to the motors, servos and actuators that drive the control surfaces. One problem with this system is the lack of 'feel' that the pilot experiences. Another is a concern over the reliability of FBW systems and the consequences of computer or electrical failure. Because of this, most FBW systems incorporate redundant computers as well as some kind of mechanical or hydraulic backup. FBW offers several important advantages. Computer control reduces the burden on a pilot and makes it possible to introduce automatic control. Another significant advantage of FBW is a significant reduction in aircraft weight which in turn reduces fuel consumption and helps to reduce undesirable CO_2 emissions. Computer control can also help to ensure that an aircraft is flown more precisely and always within its 'flight protection envelope'. The crew are thus able to cope with emergency situations without running the risk of exceeding the flight envelope or over-stressing the aircraft.

Finally, fly-by-wire technology has made it possible for aircraft manufacturers to develop 'families' of very similar aircraft. Airbus/EADS for instance has the 107-seat A318 to the 555-seat A380, with comparable flight deck designs and handling characteristics. Crew training and conversion is therefore shorter, simpler and highly cost-effective. Additionally, pilots can remain current on more than one aircraft type simultaneously.

15.5 Flight Management System (FMS)

Modern Flight Management Systems (FMS) provide advanced flight planning and navigation capability. Utilising existing combinations of avionics equipment including, GPS, VOR/DME, Inertial Reference/Navigation Systems (IRS/INS) and dead reckoning data to provide en-route, terminal and non-precision approach navigation guidance. Flight Management Systems typically provide:

- integrated automatic multi-sensor navigation
- sensor monitoring and control
- display/radar control
- moving map data display
- Steering/pitch commands to autopilot
- multiple waypoint lateral navigation (LNAV)
- optimised vertical navigation (VNAV)
- time/fuel planning and predictions based on the aircraft flight data
- data based Departure Procedures (DP)
- Standard Terminal Arrival Routes (STAR) and approaches
- integrated EFIS and radar control
- system integrity and monitoring
- flight maintenance and execution.

Figure 15.11 Typical Airbus FMC CDU

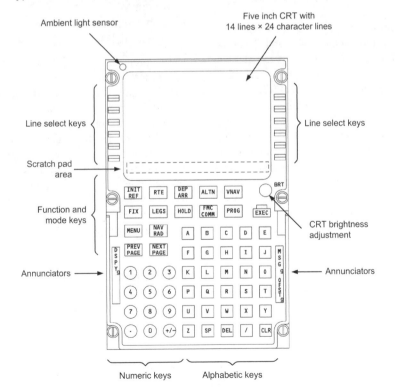

Figure 15.12 Typical Boeing FMC CDU layout

Flight Management Systems (FMS) require three main elements:

- Flight Management Computer (FMC)
- Display System (could be existing EFIS)
- Data Base Storage Unit (DBU for waypoint storage)
- Control Display Unit (CDU) with keypad.

As with other flight safety critical systems the FMS can be a 'dual fit'. Each system computes aircraft position. Cross-checks between systems ensure the validity and accuracy of flight data, offering aircrew the reassurance of dual-redundant systems for position and navigation.

Prior to departure, waypoints (including those of the origin and destination) are entered into the FMC via the Control Display Unit (CDU) to define the route (as many as 100 in some systems). FMS initialisation also involves updating the FMS with operational flight parameters such as aircraft weight and fuel load.

The FMC's **navigational database** includes airports and ground beacons and requires periodic updating (every 28 days with 13 update cycles per annum). The update is usually distributed on CDROM but requires floppy disk for installation on most aircraft (see pages 140 and 141). In flight, the FMS uses its sensor inputs to calculate such variables as fuel consumption, airspeed, position, and expected time of arrival (ETA).

Vertical flight limits are maintained by a Vertical Navigation (VNAV) system and the aircraft's autopilot system. VNAV monitors for correct speed and altitude (as determined in the flight plan) limits and ensures they are maintained at waypoints. By combining these automatic functions, a flight can be made almost entirely automatic, from initial take off to final touch down.

Key Point

The FMC requires regular updating of navigational data based on a 28 day period with 13 cycles per annum. Flight data (such as aircraft weight) must also be entered into the FMC prior to take off.

15.6 Global Positioning System (GPS)

The Global Positioning System (GPS) provides the following positional information on a world-wide basis. The information can comprise:

- latitude
- longitude
- altitude
- time
- speed.

GPS was a spin off from two experimental satellite navigation programmes carried out by the US Navy and Air Force and intended to facilitate precision aimed weapons and accurate troop deployment. Although the system is now available to the civilian market, it is still controlled and administered by the US Department of Defence. The system is highly accurate although, should the need arise, the US military can degrade the accuracy of the system to suit (up to 1,000 m). The main advantages of GPS are:

- accuracy
- global application
- signal fidelity.

Very precisely positioned orbiting satellites transmit very accurate, coded satellite position and time data. The receiver decodes this data and calculates its position relative to the satellite. If the receiver is moving then these characteristics must be included for accurate results. Using data from more than one satellite improves accuracy. The accuracy is ultimately based on very precise atomic clocks in each satellite. GPS consists of three segments; the Space Segment, the User Segment and the Control Segment. We shall briefly explain each of these in turn.

15.6.1 Space segment

The Space Segment, consists of a minimum of 24 operational satellites in six circular orbits 20,200 km (10,900 nm) above the earth at an inclination angle of 55 degrees with a 12 hour period. The satellites are spaced in orbit so that at any time a minimum of six satellites will be in view to users anywhere in the world (see Fig. 15.13). The

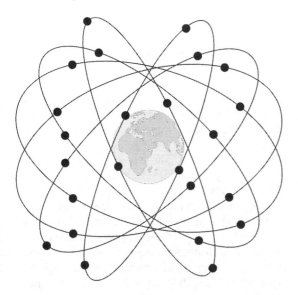

Figure 15.13 GPS uses 24 satellites in six orbital planes (there are four satellites in each plane)

Figure 15.14 A Navstar GPS satellite under construction

satellites (see Fig. 15.14) continuously broadcast position and time data to users throughout the world.

Transmission accuracy is maintained using atomic clocks accurate to within 0.1 s in 10,000 years. System accuracy is dependant on the perfect synchronisation of receiver and satellite clocks and the quality of the receiver clock. Note that using data from several satellites can help to reduce these errors.

15.6.2 User segment

The User Segment consists of the receivers, processors, and antennas that allow land, sea, or airborne operators to receive the GPS satellite broadcasts and compute their precise position, velocity and time.

15.6.3 Control segment

The Control Segment consists of a master control station in Colorado Springs, with five monitor stations and three ground antennas located throughout the world. The monitor stations track all GPS satellites in view and collect ranging information from the satellite broadcasts. The monitor stations send the information they collect from each of the satellites back to the master control station, which computes extremely precise satellite orbits. The information is then formatted into updated navigation messages for each satellite. The updated information is transmitted to each satellite via the ground antennas, which also transmit and receive satellite control and monitoring signals.

15.6.4 GPS frequencies

Each satellite transmits two carriers; L1 at 1,575.42 MHz and L2 at 1,227.60 MHz. The L1 signal is modulated by a 1.023 MHz coarse/acquisition code (C/A) that repeats every 1 ms. Although the coded signal is repetitive is appears as random and is called Pseudo Random Noise (PRN).

The L2 carrier is modulated by another apparently random coded PRN signal at 10.23 MHz. This signal is known as the Precision (P) code and it repeats every 267 days. Each satellite is allocated a unique seven day segment of this code. A third signal containing navigation data is superimposed on each already complex signal containing the satellite status, ephemeris data, clock error and tropospheric and ionospheric data for error correction. Interestingly, the US control section can 'dither' these frequencies at any time to invalidate any known codes thus making the GPS inoperative to those without knowledge of this dither coding. When this is done the code is know as the Y code and is only suitable for military/political applications.

With the exception of Y coding each GPS receiver knows the modulation codes for each satellite and it is simply a matter of decoding the received signals to identify the satellite and then removing the carrier from the received signals to extract the navigation data.

Key Point

GPS provides highly accurate navigational information. The system is based on 24 satellites in six orbits with four satellites per orbit.

Test your understanding 15.3

What FMS data needs periodic updating and when should it be updated?

Test your understanding 15.4

On what does the ultimate accuracy of GPS data depend? How can this accuracy be improved?

15.7 Inertial Reference System (IRS)

The Inertial Reference System (IRS) and Inertial Navigation System (INS) are methods of very accurate navigation that do not require any external input such as ground radio information. They are passive systems that work entirely independently of any external input, a very useful characteristic as the system cannot then be easily liable to interference. On modern aircraft the IRS is separate and supplies data to the FMC (the latter performing the navigation function). The basic IRS consists of:

- gyroscopes
- linear accelerometers
- a navigation computer

- a clock.

Gyroscopes are instruments that sense directional deviation using the characteristics of a very fast spinning mass to resist turning. Three such spinning masses are mounted orthogonally in a structure known as a **gimbal**. This frame allows the three rotating 'gyros' to maintain their direction of spin during aircraft movement thereby indicating the 'sensed' changes. In aircraft they are used to sense angles of roll, pitch, and yaw.

The **accelerometers** sense speed deviations (acceleration) along each of the axes. This three dimensional accelerometer/gyro configuration gives three orthogonal acceleration components which can be vectorially summed. Combining the gyro-sensed direction change with the summed accelerometer outputs yields the directional acceleration of the vehicle.

The system clock determines the rate the navigation computer time integrates each directional acceleration in order to obtain the aircraft's **velocity vector**. The velocity vector is then integrated with time, to obtain the **distance vector**. These steps are continuously iterated throughout the navigation process giving accurate positional information.

Crucial to the accuracy of the IRS is the initialisation of the system. This is the process of pre-flight gimbal and position alignment that gives a datum to measure all further in flight movement from.

Basically, the gyros are 'run up' to speed, the compass heading is aligned, and the latitude and longitude of the origin are entered in the computer. Any further movement of the aircraft can be calculated against this datum.

In modern avionics systems laser or optical gyros are used. These are more reliable due to containing few moving parts. In addition to this, in commercial operations, GPS data is used to further facilitate accuracy. This of course is not operationally necessary but can provide additional valuable information.

15.7.1 Gimballed systems

Gyros were initially located on a rotating platform connected to an outer housing via low friction gimbals. Accelerometers were attached to each gimballed gyro axis and thus were held in a fixed orientation. Any angular motion was sensed by the rotating platform, this maintains the platform's original orientation. Pickoffs on the gimbals measure the movement of the outer body around the steady platform and the accelerometers measure the body's acceleration in the fixed inertial axes.

Gimballed systems had a tendency to 'lock up' or 'topple' in certain violent or fast manoeuvres. Additionally they were mechanical/moving parts dependant.

The gimballed systems primary advantage is its inherently lower error. Since its three orthogonal accelerometers are held in a fixed inertial orientation, only the vertically oriented one will be measuring gravity (and therefore experiencing gravity-related errors). In strap down systems, the accelerometers all move in three axes and each experiences potential errors due to gravity.

Gimballed systems also have the advantage of simplicity of operation. The primary function of the gyro in a gimballed system is to spin and maintain a high moment of inertia, whereas strap down gyros need to actually measure the subtended angles of motion.

15.7.2 Strap down systems

With fewer moving parts, strap down systems were developed using advanced computer technologies. Progress in electronics, optics, and solid state technology have enabled very accurate reliable systems to be developed. Modern commercially available equipments take advantage of integrated circuit technologies.

Strap down systems are fixed to the aircraft structure; the gyros detecting changes in angular rate and the accelerometers detecting changes in linear rate, both with respect to the fixed axes. These three axes are a moving frame of reference as opposed to the constant inertial frame of reference in the gimballed system. The system computer uses this data to calculate the motion with respect to an inertial frame of reference in three dimensions.

The strap down system's main advantage is the simplicity of its mechanical design. Gimballed systems require complex and expensive design

for its gimbals, pickoffs, and low-friction platform connections; strap down systems are entirely fixed to the body in motion and are largely solid-state in design.

Test your understanding 15.5

1. List the components of an Inertial Reference System (IRS).

2. Explain the function of each component of an IRS in relation to the overall system.

Test your understanding 15.6

Distinguish between gimballed and strapped down gyros and state the advantages and disadvantages of each.

15.8 TCAS

The Traffic Alert Collision Avoidance System (TCAS) is a surveillance and avoidance system that alerts aircrew if any other aircraft enter a predetermined envelope of airspace around an aircraft. It is a secondary radar facility (transmits to, and receives transmissions from, other TCAS equipped aircraft).

The evaluated traffic information is displayed as symbols on the ND, the Navigation Display. Note that altitude and vertical motion information is only available if the received signal comes from a mode C or mode S transponder. Otherwise the associated symbol on the ND will have no altitude information and no vertical motion arrow.

As TCAS checks the other aircraft's relative distance permanently in short time intervals, it can therefore also calculate the other aircraft's closure rate relative to the own aircraft. The closure rate is the most important variable and indeed a very fail-safe key to a meaningful collision prediction. Complex, trigonometric calculations of flight paths and ground speeds are unnecessary and may also result in unreliable extrapolations. Note that heading or bearing information is not required to compute a TCAS alarm.

When TCAS detects that an aircraft's distance and closure rate becomes critical, it generates aural and visual annunciations for the pilots. If necessary, it also computes aural and visual pitch commands to resolve a conflict. If the other aircraft uses TCAS II as well, these pitch commands are coordinated with the other aircraft's pitch commands so that both aircraft don't 'escape' in the same direction. Even three aircraft can be coordinated.

It is important to be aware that TCAS provides only vertical guidance, no lateral guidance. TCAS also ignores performance limitations. In other words, when flying at maximum altitudes TCAS may still generate a climb command!

In addition to a transponder, various systems are required to run TCAS including:

- IRS (attitude data, vertical motion data)
- gear position sensors (as the extended gear disturbs the lower directional antenna, bearing detection of traffic flying below the own aircraft must be inhibited)
- radio altimeters (TCAS must know the radio height as the alarm logic varies with the height above ground)
- GPWS (overrides TCAS advisories during a wind shear or ground proximity warning).

When an intruder aircraft enters the protected area around an aircraft (see Fig. 15.15), TCAS triggers an alarm. The threshold of the area is defined by the time to the Closest Point of Approach (CPA) (the time-to-go is distance divided by closure rate, both combined vertically and horizontally). The protective area can be divided into two regions, one in which a Traffic Advisory (TA) message will be generated and one in which a Resolution Advisory (RA) message will be generated when an intruder appears.

The **Traffic Advisory** (TA) messages are given to the pilot in form of the word 'TRAFFIC' displayed in yellow on the ND, and the aural voice annunciation, 'traffic, traffic'. This is not the highest alert level. Its purpose is simply to call attention to a potential conflict.

TCAS triggers a TA as soon as an intruder enters the TA region. If no altitude data is available from the intruder aircraft, TCAS assumes the intruder's relative altitude is within 1200 feet. If bearing information is available, the intruder can be identified on the ND by a yellow, solid circle. Otherwise, the circle is removed and lateral distance and relative altitude with vertical motion arrow (if motion is detected) is displayed in yellow numbers under the word 'TRAFFIC'.

Resolution Advisory (RA) messages are generated at the highest TCAS alert level and they provide the pilot with aural and visual pitch commands. The pilot must then disengage the autopilot as the escape manoeuvre has to be flown manually. Any flight director commands (as well as ATC advisories may have to be ignored). The pitch command of an RA always has the highest priority. Note that, if no altitude data is available from the target, an RA will not occur. If bearing information is available, the intruder can be identified on the ND by a red, solid square. If

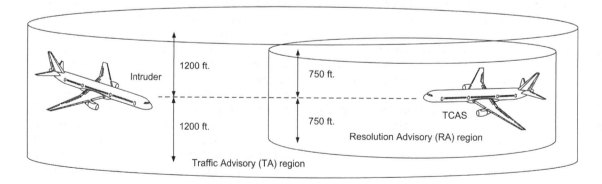

Figure 15.15 TCAS Traffic Advisory (TA) and Resolution Advisory (RA) regions

bearing information is not available, the square is removed and lateral distance and relative altitude with vertical motion arrow (if motion is detected) is displayed in red numbers under the word 'TRAFFIC'.

Although TCAS is highly regarded, it will have no angular determination warning until TCAS IV/ ACAS III is developed and proved. Several options are available including utilisation of GPS systems and Automated En-route Air Traffic (AERA) systems with up to 99.99% accuracy rates. AERA will evaluate all aircraft positions, altitude and speed. The intention is to improve the autonomy of aircraft and thereby significantly reduce ATC involvement.

Test your understanding 15.7

Give an example of a warning indication by the TCAS system. How does the pilot receive this message?

15.9 Automatic test equipment (ATE)

ATE is a dedicated ground test instrument that provides a variety of different tests and functional checks on an LRU or printed circuit card. By making a large number of simultaneous connections with the equipment under test, ATE is able to gather a large amount of data very quickly, thus avoiding the need to make a very large number of manual measurements in order to assess the functional status of an item of equipment.

ATE systems tend to be dedicated to a particular avionic system and are expensive to develop and manufacture. Because of this they tend to be only used by original equipment manufacturers (OEM) and licensed repairers.

ATE systems usually incorporate computer control with displays that indicate what further action (repair or adjustment) is necessary in order maintain the equipment. Finally, it is worth noting that, individual items of equipment may often require further detailed tests and measurements following initial diagnosis using ATE.

15.10 Built-in test equipment (BITE)

As the name implies, BITE is primarily a self-test feature built into airborne equipment as an integral fault indicator. BITE is usually designed as a signal flow type test. If the signal flow is interrupted or deviates outside accepted levels, warning alerts indicate a fault has occurred. The functions or capabilities of BITE include the following:

- real-time, critical monitoring
- continuous display presentation
- sampled recorder readout
- module and/or subassembly failure isolation
- verification of systems status
- go/no-go alarms
- quantitative displays
- degraded operation status
- percentage of functional deterioration.

The Electronic Centralised Aircraft Monitoring system (ECAM) oversees a variety of aircraft systems and also collects data on a continuous basis. While ECAM automatically warns of malfunctions, the flight crew can also manually select and monitor individual systems. Failure messages recorded by the flight crew can be followed-up by maintenance personnel by using the system test facilities on the maintenance panel in the cockpit and on the BITE facility located on each computer. The majority of these computers are located in the aircraft's avionics bay.

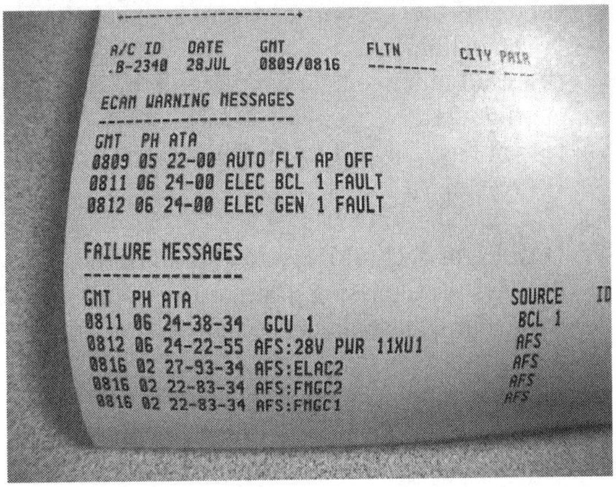

Figure 15.16 ECAM cockpit print out showing warning and failure messages

15.11 Multiple choice questions

1. The standard for ACARS is defined in:
   (a) ARINC 429
   (b) ARINC 573
   (c) ARINC 724.

2. Engine parameters are displayed on:
   (a) ECAM
   (b) EHSI
   (c) CDU.

3. A basic IRS platform has:
   (a) three accelerometers and two laser gyros
   (b) two accelerometers and three laser gyros
   (c) three accelerometers and three laser gyros.

4. The operational FMS database is:
   (a) updated once a month
   (b) is fed with information on aircraft weight before take off
   (c) needs no update information.

5. The left and right cockpit displays:
   (a) are supplied from separate symbol generators at all times
   (b) are supplied from the same symbol generator
   (c) will only be supplied from the same symbol generator when all other symbol generators have failed.

6. A single failure in a fly-by-wire system should:
   (a) cause the system to revert to mechanical operation
   (b) result in immediate intervention by the flight crew
   (c) not have any effect on the operation of the system.

7. On a flight deck EFIS system, if all of the displays were missing information from a particular source, the most likely cause would be:
   (a) the symbol generator and display
   (b) the sensor, input bus or display controller
   (c) the display controller and symbol generator.

8. A central maintenance computer provides:
   (a) ground and in-flight monitoring and testing
   (b) ground and BITE testing using a portable control panel
   (c) display of system warnings and cautions.

9. The FMS navigation database is updated:
   (a) at the request of the flight crew
   (b) during pre-flight checks
   (c) every 28 days.

10. The sweep voltage waveform used on an electromagnetic CRT is:
    (a) trapezoidal
    (b) sinusoidal
    (c) sawtooth.

11. In an EFIS with three symbol generators, what is the purpose of the third symbol generator?
    (a) Comparison with the pilot's symbol generator
    (b) Standby in case of failure
    (c) To provide outputs for ECAM.

12. The ACARS system uses channels in the:
    (a) HF spectrum
    (b) VHF spectrum
    (c) UHF spectrum.

13. If one EICAS CRT fails:
    (a) the remaining CRT will display primary EICAS data
    (b) the FMS CDU will display the failed CRT data
    (c) the standby CRT will automatically take over.

14. A method used in modern aircraft for reporting in-flight faults to an engineering and monitoring ground station is:
    (a) TCAS II
    (b) ACARS
    (c) GPS.

Appendix 1 — Abbreviations and acronyms

Abbrev.	Meaning
ACARS	Aircraft Communication Addressing and Reporting System
ACAS	Airborne Collision Avoidance System
ACM	Aircraft Condition Monitoring
ACMS	Aircraft Condition Monitoring System
ACR	Avionics Computing Resource
ACU	Aviation Computing Unit
ADAPT	Air Traffic Management Data Acquisition Processing and Transfer
ADC	Analog to Digital Converter
ADF	Automatic Direction Finder
ADI	Attitude Director Indicator
ADIRS	Air Data/Inertial Reference System
ADIRU	Air Data Inertial Reference Unit
ADM	Air Data Module
ADR	Air Data Reference
ADS	Air Data System
AEEC	Airlines Electronic Engineering Committee
AFC	Automatic Frequency Control
AFCS	Auto Flight Control System (Autopilot)
AFDX	Avionics Full Duplex
AFIS	Airborne Flight Information Service
AFMS	Advanced Flight Management System
AFS	Automatic Flight System (Autopilot)
AGL	Above Ground Level
AHRS	Attitude/Heading Reference System
AHS	Attitude Heading System
AI	Airbus Industries
AIAA	American Institute of Aeronautics and Astronautics
AIDS	Aircraft Integrated Data System
AIMS	Airplane Information Management System
AIS	Aeronautical Information System
AIV	Anti-Icing Valve
ALT	Altitude
ALU	Arithmetic Logic Unit
AM	Amplitude Modulation
AMD	Advisory Map Display
AMI	Airline Modifiable Information
AMLCD	Active Matrix Liquid Crystal Display
ANSI	American National Standards Institute
ANSIR	Advanced Navigation System Inertial Reference
AOA	Angle of Attack
AP	Autopilot
APATSI	Airport Air Traffic System Interface
API	Application Programming Interface
APIC	Advanced Programmable Interrupt Controller
APM	Advanced Power Management
APP	Approach
APR	Auxiliary Power Reserve
APU	Auxiliary Power Unit
ARINC	Aeronautical Radio Incorporated
ARTAS	Advanced Radar Tracker and Server
ARTS	Automated Radar Terminal System
ASAAC	Allied Standard Avionics Architecture Council
ASCB	Aircraft System Common Bus
ASCII	American Standard Code for Information Interchange
ASI	Air Speed Indicator
ASIC	Application Specific Integrated Circuit
ASR	Aerodrome Surveillance Radar
ATA	Actual Time of Arrival
ATC	Air Traffic Control
ATE	Automatic Test Equipment
ATFM	Air Traffic Flow Management
ATI	Air Transport Indicator
ATLAS	Abbreviated Test Language for Avionics Systems
ATM	Air traffic management
ATN	Aeronautical Telecommunications Network
ATR	Air Transportable Racking
ATS	Air Traffic Services
ATSU	Air Traffic Services Unit
AVLAN	Avionics Local Area Network
AWLU	Aircraft Wireless Local Area Network Unit
BC	Bus Controller
BCD	Binary Coded Decimal
BGA	Ball Grid Array
BGW	Basic Gross Weight
BIC	Backplane Interface Controller
BIOS	Basic Input/Output System
BIST	Built-In Self-Test
BIT	Built-in Test
BITE	Built-in Test Equipment
BIU	Bus Interface Unit
BNR	Binary Numeric data

BP	Base Pointer
BPRZ	Bipolar Return to Zero
BPS	Bits Per Second
BPV	Bypass Valve
C/PDLC	Controller/Pilot Data Link Communications
CAA	Civil Aviation Authority
CABLAN	Cabin Local Area Network
CADC	Central Air Data Computer
CAI	Computer Aided Instruction
CAN	Controller Area Network
CAS	Collision Avoidance System
CAS	Crew Alerting System
CAT	Clear-Air-Turbulence
CCA	Circuit Card Assembly
CDDI	Copper Distributed Data Interface
CDI	Course Deviation Indicator
CDROM	Compact Disk Read-Only Memory
CDS	Common Display System
CDU	Control Display Unit
CEATS	Central European Air Traffic Service
CFDS	Central Fault Display System
CH	Compass Heading
CIDIN	Common ICAO Data Interchange Network
CIDS	Cabin Intercommunication Data System
CISC	Complex Instruction Set Computer
CLK	Clock
CMC	Central Maintenance Computer
CMOS	Complementary Metal Oxide Semiconductor
CMP	Configuration Management Plan
CMS	Centralized Maintenance System
CMU	Communications Management Unit
COMPAS	Computer Orientated Metering, Planning and Advisory System
COTS	Commercial off-the-shelf
CPA	Closest Point of Approach
CPGA	Ceramic Pin Grid Array
CPM	Core Processing Module
CPU	Central Processing Unit
CRC	Cyclic Redundancy Check
CRM	Crew Resource Management
CRT	Cathode Ray Tube
CS	Code Segment
CTO	Central Technical Operations
CVR	Cockpit Voice Recorder
DA	Drift Angle
DAC	Digital to Analog Converter
DADC	Digital Air Data Computer
DATAC	Digital Autonomous Terminal Access Communications System
DBRITE	Digital Bright Radar Indicator Tower Equipment
DCU	Data Concentrator Unit
DDR	Digital Data Recorder

DECU	Digital Engine Control
DEOS	Digital Engine Operating System
DEU	Digital Electronics Unit
DFDAU	Digital Flight Data Acquisition Unit
DFDR	Digital Flight Data Recorder
DFGC	Digital Flight Guidance Computer
DFGS	Digital Flight Guidance System
DFLD	Database Field Loadable Data
DG	Directional Gyro
DGPS	Differential Global Positioning System
DH	Decision Height
DI	Destination Index
DIL	Dual In-line
DIMM	Dual In-line Memory Module
DIP	Dual In-line Package
DITS	Digital Information Transfer System
DLS	DME-based Landing System
DMA	Direct Memory Access
DME	Distance Measuring Equipment
DMEP	Data Management Entry Panel
DO	Design Organisation
DP	Departure Procedures
DP	Decimal Point
DPM	Data Position Module
DPU	Display Processor Unit
DRAM	Dynamic Random Access Memory
DS	Data Segment
DSCS	Door and Slide Control System
DSP	Display Select Panel
DSP	Digital Signal Processing
DTOP	Dual Threshold Operation
DU	Display Units
DUATS	Direct User Access Terminal System
EADI	Electronic Attitude Director Indicator
EARTS	En-route Automated Radar Tracking System
EAS	Express Air System
EASA	European Aviation Safety Agency
EAT	Expected Approach Time
EATMS	Enhanced Air Traffic Management System
ECAM	Electronic Centralized Aircraft Monitoring
ECB	Electronic Control Box
ECC	Error Checking and Correcting
ECM	Electronic Countermeasures
ECP	Extended Capabilities Port
ECS	Environmental Control System
ECU	Electronic Control Unit
EEC	Electronic Engine Control
EEPROM	Electrically Erasable Programmable Read-Only Memory
EFCS	Electronic Flight Control System
EFIS	Electronic Flight Instrument System
EGPWS	Enhanced Ground Proximity Warning System
EHSI	Electronic Horizontal Situation Indicator

EIA	Electronic Industries Association
EICAS	Engine Indication and Crew Alerting Systems
EIDE	Enhanced Integrated Drive Electronics
EIS	Electronic Instrument System
EL	Elevation-Station
ELAC	Elevator and Aileron Computer
ELS	Electronic Library System
EM	Emulate Processor
EMC	Electromagnetic Compatibility
EMI	Electromagnetic Interference
EOC	End of Conversion
EPP	Enhanced Parallel Port
EPROM	Erasable Programmable Read-Only Memory
EROPS	Extended Range Operations
ERU	Electronic Routing Unit
ES	Extra Segment
ESD	Electrostatic Discharge
ESD	Electrostatic Sensitive Device
ETA	Estimated Time of Arrival
ETOPS	Extended Range Twin-engine Operations
EU	Execution Unit
FAA	Federal Aviation Administration
FAC	Flight Augmentation Computer
FADEC	Full Authority Digital Engine Control
FAMIS	Full Aircraft Management/Inertial System
FAR	Federal Aviation Regulations
FBL	Fly-by-light
FBW	Fly-by-wire
FC	Flight Control
FCC	Flight Control Computer
FCC	Federal Communications Commission
FCGC	Flight Control and Guidance Computer
FCS	Flight Control System
FCU	Flight Control Unit
FD	Flight Director
FDAU	Flight Data Acquisition Unit
FDC	Fight Director Computer
FDD	Floppy Disk Drive
FDDI	Fibre Distributed Data Interface
FDMU	Flight Data Management Unit
FDR	Flight Data Recorder
FDS	Flight Director System
FET	Field Effect Transistor
FFS	Full Flight Simulator
FG	Flight Guidance
FGC	Flight Guidance Computer
FGI	Flight guidance by digital Ground Image
FGS	Flight Guidance System
FIFO	First-In First-Out
FIR	Flight Information Region
FIS	Flight Information System
FL	Flight Level
FLIR	Forward Looking Infrared

FLS	Field Loadable Software
FM	Flight Management
FMC	Flight Management Computer
FMCDU	Flight Management Control and Display Unit
FMCS	Flight Management Computer System
FMGC	Flight Management Guidance Computer
FMS	Flight Management System
FOG	Fibre Optic Gyros
FSC	Fuel System Controller
G	Giga (10^9 multiplier)
G/S	Glide slope
GA	General Aviation
GBST	Ground-based Software Tool
GG	Graphics Generator
GHz	Gigahertz (10^9 Hz)
GMT	Greenwich Mean Time
GND	Ground
GNSS	Global Navigation Satellite System
GPM	Ground Position Module
GPS	Global Positioning System
GPWS	Ground Proximity Warning System
GS	Ground Speed
HALS	High Approach Landing System
HDD	Head Down Display
HDG	Heading
HIRF	High-energy Radiated Field/High-intensity Radiated Field
Hex	Hexadecimal
HFDS	Head-up Flight Display System
HGS	Head-up Guidance System
HIRF	High-intensity Radiated Field
HIRL	High-intensity Runway Lights
HM	Health Monitoring
HMCDU	Hybrid Multifunction Control Display Unit
HPA	High Power Amplifier
HSI	Horizontal Situation Indicator
HUD	Head-Up Display
HUGS	Head-Up Guidance System
HW	Hardware
Hz	Hertz (cycles per second)
I/O	Input/output
IAC	Integrated Avionics Computer
IAPS	Integrated Avionics Processing System
IAS	Indicated Air Speed
IBC	Individual Blade Control
IBR	Integrated Bladed Rotor
IDE	Integrated Drive Electronics
IFE	In-Flight Entertainment
IFF	Identification, Friend or Foe
IFOG	Interferometric Fibre Optic Gyro
IFPS	International Flight Plan Processing System
IFR	Instrument Flight Rules

IHF	Integrated Human Interface Function	LSB	Least Significant Bit
IHUMS	Integrated Health and Usage Monitoring System	LSD	Least Significant Digit
		LSI	Large Scale Integration
ILS	Instrument Landing System	LSS	Lightning Sensor System
IMA	Integrated Modular Avionics		
IMU	Inertial Measurement Unit	M	Mega (10^6 multiplier)
INS	Inertial Navigation System	MASI	Mach Airspeed Indicator
IOAPIC	Input/Output Advanced Programmable Input Controller	MAU	Modular Avionics Unit
		MCDU	Microprocessor Controlled Display Units
IOM	Input/Output Module	MCP	Mode Control Panel
IP	Internet Protocol	MCU	Modular Component Unit
IPC	Instructions Per Cycle	MDAU	Maintenance Data Acquisition Unit
IPR	Intellectual Property Right	MEL	Minimum Equipment List
IPX/SPX	Inter-network Packet Exchange/Sequential Packet Exchange	MFD	Multi-function Flight Display
		MFDS	Multi-function Display System
IR	Infra-Red	MHRS	Magnetic Heading Reference System
IRQ	Interrupt Request	MHX	Main Heat Exchanger
IRS	Inertial Reference System	MHz	Megahertz (10^6 Hz)
IRU	Inertial Reference Unit	MIPS	Millions of instructions per second
ISA	Industry Standard Architecture	MLS	Main Sea Level
ISA	Inertial Sensor Assembly	MLW	Maximum Landing Weight
ISAS	Integrated Situational Awareness System	MMI	Man Machine Interface
ISDB	Integrated Signal Data Base	MMM	Mass Memory Module
ISDU	Inertial System Display Unit	MMS	Mission Management System
ISO	International Standards Organization	MNPS	Minimum Navigation Performance Specification
IVSI	Instantaneous Vertical Speed Indicator		
IWF	Integrated Warning Function	MOS	Metal Oxide Semiconductor
		MOSFET	Metal Oxide Semiconductor Field Effect Transistor
JAA	Joint Airworthiness Authority		
JAR	Joint Airworthiness Requirement	MP	Monitor Processor
JEDEC	Joint Electron Device Engineering Council	MPT	Memory Protocol Translator
		MPU	Multifunction Processor Unit
K	Kilo (10^3 multiplier)	MRC	Modular Radio Cabinets
kHz	Kilohertz (10^3 Hz)	MRO	Maintenance/Repair/Overhaul
KIAS	Indicated Airspeed in Knots	MSB	Most Significant Bit
Knot	Nautical miles/hour	MSD	Most Significant Digit
KT	Knots	MSI	Medium Scale Integration
		MSU	Mode Select Unit
LAAS	Local Area Augmentation System	MSW	Machine Status Word
LAN	Local Area Network	MT	Maintenance Terminal
LATAN	Low-Altitude Terrain-Aided Navigation	MTBF	Mean Time Between Failure
LBA	Logical Block Addressing	MTBO	Mean Time Between Overhaul
LCC	Leadless Chip Carrier	MTC	Mission and Traffic Control systems
LCD	Liquid Crystal Display	MTOW	Maximum Takeoff Weight
LDU	Lamp Driver Unit		
LED	Light-emitting Diode	NASA	National Aeronautics and Space Administration
LFC	Laminar Flow Control		
LGC	Landing Gear Control	NCD	No Computed Data
LIDAR	Light Radar	ND	Navigation Display
LNAV	Lateral Navigation	NDB	Navigation Data Base
LOC	Localizer	NetBIOS	Network Basic Input Output System
LORADS	Long Range Radar and Display System	NIC	Network Interface Controller
LRM	Line Replaceable Module	NMI	Non-Maskable Interrupt
LRU	Line Replaceable Unit	NMS	Navigation Management System
LRU	Least Recently Used	NMU	Navigation Management Unit
LSAP	Loadable Aircraft Software Part	NRZ	Non-Return to Zero

NVM	Non-Volatile Memory		RISC	Reduced Instruction Set Computer
NVS	Noise and Vibration Suppression		RLG	Ring Laser Gyro
			RMI	Radio Magnetic Indicator
OAT	Outside Air Temperature		RNAV	Area Navigation
OBI	Omni Bearing Indicator		RNP	Required navigation performance
OBS	Omni Bearing Selector		ROM	Read-Only Memory
ODS	Operations Display System			
OEI	One Engine Inoperative		SAARU	Secondary Attitude/Air Data Reference Unit
OEM	Original Equipment Manufacturer			
OLDI	On-Line Data Interchange		SAR	Successive Approximation Register
OMS	On-board Maintenance System		SAT	Static Air Temperature
OS	Operating System		SATCOM	Satellite communications
OSS	Option Selectable Software		SC	Start Conversion
OTP	One-time Programmable		SCMP	Software Configuration Management Plan
			SDD	System Definition Document
PA	Physical Address		SDI	Source/Destination Identifier
PAC	PCI AGP Controller		SDIP	Shrink Dual In-line Package
PAPI	Precision Approach Path Indicator		SEC	Secondary
PBGA	Plastic Ball Grid Array		SI	Source Index
PCA	Preconditioned Air System		SIL	Single In-line
PCB	Printed Circuit Board		SIP	Single In-line Package
PCC	Purser Communication Centre		SMART	Standard Modular Avionics Repair and Test
PCHK	Parity Check(ing)			
PCI	Peripheral Component Interconnect		SMD	Surface Mounted Device
PCMCIA	Personal Computer Memory Card International Association		SMT	Surface mount technology
			SNMP	Simple Network Management Protocol
PDL	Portable Data Loader		SO	Small Outline
PFD	Primary Flight Display		SOIC	Small Outline Integrated Circuit
PGA	Pin Grid Array		SOJ	Small Outline J-lead
PIC	Pilot In Command		SOP	Small Outline Package
PLCC	Plastic Leadless Chip Carrier		SP	Stack Pointer
PM	Protected Mode		SPDA	Secondary Power Distribution Assembly
PMAT	Portable Maintenance Access Terminal		SPDT	Single Pole Double Throw
PMO	Program Management Organization		SRAM	Synchronous Random Access Memory
PMS	Performance Management System		SRD	System Requirement Document
PMSM	Power Management State Machine		SROM	Serial Read Only Memory
PNF	Pilot Non Flying		SS	Stack Segment
POST	Power-On Self-test		SSI	Small Scale Integration
PP	Pre-Processor		SSM	Sign/Status Matrix
PQFP	Plastic Quad Flat Package		SSOP	Shrink Small Outline Package
PRF	Pulse Repetition Frequency		STAR	Standard Terminal Arrival Routes
PRI	Primary		STP	Shielded Twisted Pair
PROM	Programmable Read-Only Memory		SW	Software
PSM	Power Supply Module			
PSU	Bypass Switch Unit		TA	Traffic Advisory
PWB	Printed Wiring Board		TACAN	Tactical Air Navigation
			TAS	True Air Speed
QFP	Quad flat pack		TAT	Total air temperature
			TAWS	Terrain Awareness Warning System
R/T	Receiver/transmitter		TBO	Time Between Overhaul
RA	Resolution Advisory		TCAS	Traffic Alert Collision Avoidance System
RA	Radio Altitude		TRU	Transformer Rectifier Unit
RAS	Row Address Select		TS	Task switched
RAM	Random Access Memory		TSOP	Thin small package outline
RF	Radio Frequency		TTL	Transistor-transistor logic
RIMM	RAM Bus In-line Memory Module		TTP	Time Triggered Protocol

TWDL	Two-way data link

UDP	User Datagram Protocol
UHF	Ultra High Frequency
ULSI	Ultra Large Scale Integration
UMS	User Modifiable Software
USB	Universal Serial Bus
UTP	Unshielded Twisted Pair
UV	Ultra-violet
UVPROM	Ultra- violet Programmable Read-only Memory

VAC	Volts, Alternating Current
VAS	Virtual Address Space
VDC	Volts, Direct Current
VFR	Visual Flight Rules
VG	Vertical Gyro
VGA	Video Graphics Adapter
VHF	Very High Frequency
VHSIC	Very High Speed Integrated Circuit
VIA	Versatile Integrated Avionics
VLF	Very Low Frequency
VLSI	Very Large Scale Integration
VME	Versa Module Eurocard
VNAV	Vertical Navigation
VOR	Very High Frequency Omni Range
VPA	Virtual Page Address
VSI	Vertical Speed Indicator

WAAC	Wide Angle Airborne Camera
WE	Write Enable
WORM	Write-Once Read-Many
WX	Weather
WXP	Weather Radar Panel
WXR	Weather Radar

ZIF	Zero Insertion Force
ZIP	Zig-zag In-line Package

Appendix 2

Revision papers

These revision papers are designed to provide you with practice for examinations. The questions are typical of those used in CAA and other examinations. Each paper has 20 questions and each should be completed in 25 minutes. Calculators and other electronic aids must not be used.

Revision Paper 1

1. Typical displays on an EHSI are:
   (a) engine indications and warnings
   (b) VOR, heading and track
   (c) VOR, altitude and rate of climb.

2. EMI can be conveyed from a source to a receiver by:
   (a) conduction only
   (b) radiation only
   (c) conduction and radiation.

3. The hexadecimal number A7 is equivalent to:
   (a) 107
   (b) 147
   (c) 167.

4. A NAND gate with its output inverted has the same logical function as:
   (a) an OR gate
   (b) a NOR gate
   (c) an AND gate.

5. Which of the following types of ADC is the fastest?
   (a) flash type
   (b) dual slope type
   (c) successive approximation type.

6. Noise generated by a switching circuit is worse when:
   (a) switching is fast and current is low
   (b) switching is slow and current is high
   (c) switching is fast and current is high.

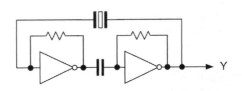

Figure A2.1 See Paper 1, Question 7

7. The circuit shown in Figure A2.1 is:
   (a) a clock generator
   (b) a ramp generator
   (c) a sine wave oscillator.

8. Which computer bus is used to specify memory locations:
   (a) address bus
   (b) control bus
   (c) data bus.

9. Figure A2.2 shows a bus interface. At which point will the data appear in parallel format?
   (a) X
   (b) Y
   (c) Z.

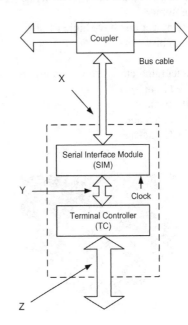

Figure A2.2 See Paper 1, Question 9

10. Which aircraft standard applies to fibre optic networks?
    (a) ARINC 429
    (b) ARINC 573
    (c) ARINC 636.

11. GPS uses:
    (a) 12 satellites in three orbital planes
    (b) 24 satellites in six orbital planes
    (c) 48 satellites in 12 orbital planes.

12. A significant advantage of fibre optic networks in modern passenger aircraft is:
    (a) reduced weight
    (b) lower installation costs
    (c) ease of maintenance.

13. Adjacent blue and red phosphors are illuminated on the screen of a colour CRT. The resulting colour produced will be:
    (a) cyan
    (b) yellow
    (c) magenta.

14. When compared with TTL devices, CMOS logic devices use:
    (a) less power and have lower noise immunity
    (b) more power and have lower noise immunity
    (c) less power and have higher noise immunity.

15. The integrated circuit shown in Figure A2.3 is supplied in a:
    (a) DIL package
    (b) PGA package
    (c) PLCC package.

Figure A2.4 See Paper 1, Question 16

16. The purpose of the items shown in Figure A2.4 is:
    (a) to ensure electrical integrity of the aircraft ground
    (b) to provide a return path for AC power supplies
    (c) to provide shielding for nearby data cables.

17. An R-2R ADC has values of resistance:
    (a) whose precision must be accurate
    (b) whose relative precision must be accurate
    (c) whose precision are unimportant.

18. An ALU is an example of:
    (a) LSI technology
    (b) MSI technology
    (c) SSI technology.

19. Which one of the following devices is most susceptible to damage from stray static charges?
    (a) a TTL logic gate
    (b) a MOSFET transistor
    (c) a silicon controlled rectifier.

20. A CMOS logic gate is operated from a 15 V supply. If a voltage of 9 V is measured at the input of the gate this would be equivalent to:
    (a) a logic 0 input
    (b) a logic 1 input
    (c) an indeterminate input.

Figure A2.3 See Paper 1, Question 15

Revision Paper 2

1. When expressed in hexadecimal an octal value of 54 is the same as:
   (a) 2C
   (b) 2F
   (c) 4F.

2. A heading reference of 320° in a word label format would be appear as:
   (a) 000010011
   (b) 011010000
   (c) 001000111.

3. The cone of acceptance of an optical fibre and a light source is measured between:
   (a) the two outer angles
   (b) the diameter of the core
   (c) the longitudinal axis of the core and the outer angle.

4. The flight director system receives information from:
   (a) attitude gyro, engines, radar altimeter
   (b) attitude gyro, VOR/localizer, compass system
   (c) compass system, weather radar, radar altimeter.

5. In the logic circuit shown in Figure A2.5, A = 1, B = 1, C = 1 and D = 1. The outputs X and Y will then be:
   (a) X = 0 and Y = 0
   (b) X = 0 and Y = 1
   (c) X = 1 and Y = 0.

Figure A2.6 See Paper 2, Question 6

6. The logic device shown in Figure A2.6 is:
   (a) a standard TTL device
   (b) a low-speed CMOS logic device
   (c) a low-power Schottky TTL device.

7. The method of data encoding used in an ARINC 429 bus is known as:
   (a) bipolar return to zero
   (b) universal asynchronous serial
   (c) synchronous binary coded decimal.

8. The integrated circuits shown in Figure A2.7 are:
   (a) RAM devices
   (b) UV-EPROM devices
   (c) flash memories.

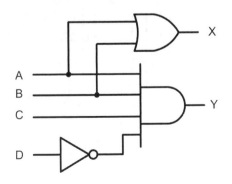

Figure A2.5 See Paper 2, Question 5

Figure A2.7 See Paper 2, Question 8

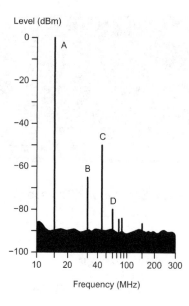

Figure A2.8 See Paper 2, Question 9

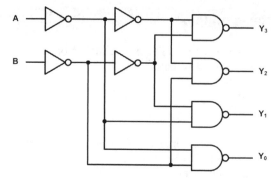

Figure A2.9 See Paper 2, Question 12

9. The output of a spectrum analyser is shown in Figure A2.8. Which of the features shown is the second harmonic?
(a) A
(b) B
(c) C.

10. A level C software classification is one in which a failure could result in:
(a) aircraft loss
(b) fatal injuries to passengers or crew
(c) minor injuries to passengers or crew.

11. The logic gate arrangement shown in Figure A2.9 will produce the:
(a) OR logic function
(b) AND logic function
(c) NAND logic function.

12. The logic circuit arrangement shown in Figure A2.10 is:
(a) a two to four line encoder
(b) a two to four line decoder
(c) a two to two line multiplexer.

13. Data is converted from serial to parallel and parallel to serial by means of:
(a) a parallel register
(b) a shift register
(c) a synchronous counter.

14. The logic device shown in Figure A2.10 is:
(a) an four to three line encoder
(b) an eight to three line encoder
(c) a four to eight line multiplexer.

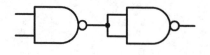

Figure A2.9 See Paper 2, Question 11

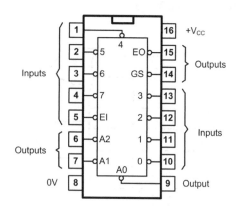

Figure A2.10 See Paper 2, Question 14

15. A group of bits transmitted at the same time is referred to as:
    (a) a clock signal
    (b) parallel data
    (c) serial data.

16. A typical characteristic of CMOS logic is:
    (a) low power dissipation
    (b) high voltage handling
    (c) high power dissipation.

17. A character generator ROM is used in conjunction with:
    (a) a GPS receiver
    (b) a CRT controller
    (c) a seven-segment indicator.

18. The feature marked R in Figure A2.11 is:
    (a) a bus cable
    (b) a stub cable
    (c) a fibre optic connection.

19. The main constituents of a computer are:
    (a) ALU, accumulator, registers
    (b) CPU, ROM, RAM, I/O
    (c) CPU, ALU, accumulator, clock.

20. In order to produce an AND gate from a NOR gate you would need to:
    (a) invert the output
    (b) invert each of the inputs
    (c) invert the output and each of the inputs.

Revision Paper 3

1   The EADI displays:
    (a) pitch and roll attitudes
    (b) heading and weather radar
    (c) pitch, roll and waypoints.

2. On an EHSI in weather radar mode, a severe storm would be shown as:
    (a) orange areas with black or yellow surrounds
    (b) blue areas with white background
    (c) red areas with black surrounds.

3. Which feature Figure A2.11 is used to prevent problems with reflection and mismatch?
    (a) P
    (b) Q
    (c) S.

4. The number of individual addresses that can be addressed by a processor that has a 16-bit address bus is:
    (a) 16,384
    (b) 32,768
    (c) 65,536.

5. In order to produce a NAND gate from an AND gate you would need to:
    (a) invert the output
    (b) invert each of the inputs
    (c) invert the output and each of the inputs.

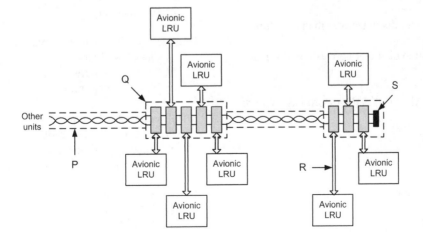

Figure A2.11 See Paper 2, Question 18 and Paper 3 Question 3

A	B	Y
0	0	1
0	1	1
1	0	1
1	1	0

Figure A2.12 See Paper 3, Question 6

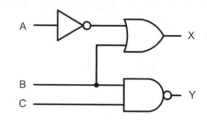

Figure A2.13 See Paper 3, Question 12

6. The truth table shown in Figure A2.12 is for:
   (a) an OR gate
   (b) a NOR gate
   (c) a NAND gate.

7. If one EICAS display fails:
   (a) the remaining CRT will display primary EICAS data
   (b) the FMS CDU will display the failed CRT data
   (c) the standby CRT will automatically take over.

8. The typical wavelength of light in an optical fibre is:
   (a) 575 nm
   (b) 635 nm
   (c) 1300 nm.

9. A monomode optical fibre has:
   (a) a larger diameter core than a multimode fibre
   (b) a smaller diameter core than a multimode fibre
   (c) the same diameter core as a multimode fibre.

10. A basic ARINC 429 data word has a length of:
    (a) 8 bits
    (b) 16 bits
    (c) 32 bits.

11. Electronic engine control software is an example of:
    (a) DFLD
    (b) LSAP
    (c) OSS.

12. Figure A2.13 shows a logic gate arrangement. In order for X and Y to both be at logic 1:
    (a) all inputs should be at logic 0
    (b) all inputs should be at logic 1
    (c) inputs B and C should be at logic 1.

13. The BCD data field of ARINC 429 is contained within bits:
    (a) 1 to 8
    (b) 1 to 10
    (c) 11 to 29.

14. To create a bidirectional communications link within an ARINC 429 system:
    (a) only one data bus connection is required
    (b) two data bus connections are required
    (c) four data bus connections are required.

15. The binary number 10100111 is equivalent to:
    (a) 107 decimal
    (b) 147 decimal
    (c) 167 decimal.

16. Weather radar is an example of:
    (a) Class B software
    (b) Class C software
    (c) Class D software.

17. GPS consists of the following segments:
    (a) space, user and control
    (b) latitude, longitude, time
    (c) carrier, modulation, encoding.

18. The octal number 127 is equivalent to:
    (a) 1001011 binary
    (b) 1101001 binary
    (c) 1010111 binary.

19. ACARS signals are transmitted at:
  (a) HF in the range 11 MHz to 14 MHz
  (b) VHF in the range 118 MHz to 136 MHz
  (c) UHF in the range 405 MHz to 420 MHz.

20. An advantage of a dual-slope ADC is:
  (a) a fast conversion time
  (b) an inherent ability to reject noise
  (c) the ability to operate without a clock.

Revision Paper 4

1. Fibre optic data cables are:
  (a) bidirectional
  (b) unidirectional
  (c) simplex.

2. In a static memory circuit:
  (a) the memory is retained indefinitely
  (b) the memory is lost as soon as power is removed
  (c) the memory needs to be refreshed constantly, even when power is on.

3. An example of a 'write once and read many' mass storage device is:
  (a) a CDROM
  (b) a floppy disk
  (c) a dynamic RAM.

4. The Q output of an R-S bistable is at logic 0. This means that the bistable is:
  (a) set
  (b) reset
  (c) in a high-impedance state

5. In an analogue to digital converter, the input voltage
  (a) can vary continuously in voltage level
  (b) can have two possible voltage levels
  (c) can only have one fixed voltage level.

6. Electromagnetic compatibility is achieved by
  (a) using smoothed and well regulated d.c. supplies
  (b) coating all equipment enclosures with conductive paint
  (c) shielding, screening, earthing, bonding and interference filters.

Figure A2.14 See Paper 4, Question 7

7. The display shown in Figure A2.14 is:
  (a) EICAS
  (b) EFIS PFD
  (c) FMS CDU.

8. An ARINC 629 data bus cable uses:
  (a) a pair of twisted wires
  (b) a single wire and ground
  (c) a coaxial cable with a single screened conductor.

9. Losses in optical fibre data cables increase when:
  (a) a low data rate is used
  (b) the cable is kept as short as possible
  (c) the cable is bent round a small radius.

10. Fibre optic cables use:
  (a) a reflective cladding
  (b) a refractive cladding
  (c) an opaque cladding.

11. An LCD uses what type of power supply?
  (a) AC voltage
  (b) current limited DC current
  (c) voltage limited DC voltage.

12. The hexadecimal equivalent of binary 10110011 is:
  (a) 31
  (b) B3
  (c) 113.

A	B	Y
0	0	1
0	1	0
1	0	0
1	1	0

Figure A2.15 See Paper 4, Question 13

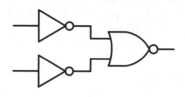

Figure A2.16 See Paper 4, Question 13

13. The truth table shown in Figure A2.15 is for:
    (a) a NOR gate
    (b) a NAND gate
    (c) an exclusive-OR gate.

14. The operational data base of the FMS may
    have to be modified in flight:
    (a) by the pilot
    (b) automatically by the DADC
    (c) It cannot be modified in flight.

15. On an EFIS system the weather radar is
    displayed on:
    (a) the EADI
    (b) the EHSI
    (c) the FMC CDU.

16. A method used in modern aircraft for
    reporting in flight faults to an engineering and
    monitoring ground station is
    (a) TCAS II
    (b) ACARS
    (c) TAWS.

17. When compared with an LCD, a CRT display
    offers the advantage that:
    (a) the resolution and focus is better
    (b) it can be viewed over a large angle
    (c) it uses low voltages and is more energy
        efficient.

18. The logic arrangement shown in Figure A2.16
    is equivalent to:
    (a) an OR gate
    (b) an AND gate
    (c) a NAND gate.

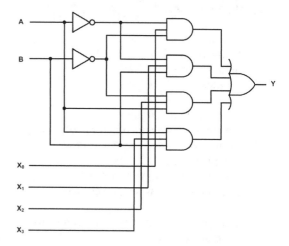

Figure A2.17 See Paper 4, Question 19

19. The logic arrangement shown in Figure A2.17
    acts as:
    (a) a decoder
    (b) a multiplexer
    (c) a shift register

20. The pins on a DIL packaged integrated circuit
    are numbered:
    (a) clockwise starting at pin-1
    (b) clockwise starting at pin-14
    (c) anticlockwise starting at pin-1.

Answers

9. c
10. c
11. b
12. a
13. a.

Answers to review questions

Chapter 1 (page 11)

1. b
2. a
3. c
4. a
5. a
6. b
7. b
8. b
9. a
10. a
11. a
12. a
13. b
14. b
15. a
16. c
17. c
18. b
19. a
20. a
21. c
22. b
23. b
24. a
25. a
26. c
27. b
28. c.

Chapter 2 (page 22)

1. b
2. c
3. a
4. b
5. a
6. c
7. b
8. c

Chapter 3 (page 32)

1. a
2. b
3. b
4. b
5. c
6. c
7. b
8. c
9. c
10. a
11. b
12. c
13. b.

Chapter 4 (page 44)

1. b
2. c
3. b
4. a
5. a
6. c
7. a
8. c
9. a
10. b
11. b
12. c
13. c
14. a
15. c
16. b
17. c.

Chapter 5 (page 61)

1. b
2. b
3. c
4. b
5. a
6. a
7. c
8. a

9.   a
10.   b.

Chapter 6 (page 77)

1.   b
2.   a
3.   b
4.   a
5.   b
6.   b
7.   b
8.   c
9.   a
10.   b
11.   c
12.   a
13.   b
14.   c
15.   c
16.   a
17.   a.

Chapter 7 (page 95)

1.   a
2.   c
3.   b
4.   b
5.   c
6.   a
7.   c
8.   a
9.   a
10.   c
11.   b
12.   b
13.   a
14.   b
15.   a
16.   c
17.   b
18.   c
19.   b
20.   b
21.   a
22.   c.

Chapter 8 (page 101)

1.   a
2.   a

3.   b
4.   c
5.   a
6.   b
7.   a
8.   c
9.   b
10.   a
11.   b
12.   c
13.   a
14.   b
15.   b.

Chapter 9 (page 110)

1.   c
2.   a
3.   c
4.   a
5.   b
6.   b
7.   a
8.   b.

Chapter 10 (page 117)

1.   b
2.   c
3.   a
4.   c
5.   a
6.   c
7.   b
8.   c
9.   a
10.   a
11.   c
12.   b
13.   b
14.   b
15.   c
16.   a
17.   c
18.   c
19.   b
20.   a.

Chapter 11 (page 130)

1.   c
2.   b

3.   a
4.   a
5.   b
6.   c
7.   b
8.   a
9.   c
10. b
11. c
12. c
13. b
14. a
15. c
16. a.

Chapter 12 (page 136)

1.   c
2.   c
3.   c
4.   c
5.   c
6.   a
7.   a
8.   a.

Chapter 13 (page 144)

1.   c
2.   b
3.   c
4.   c
5.   b
6.   a
7.   b.

Chapter 14 (page 156)

1.   a
2.   c
3.   c
4.   c
5.   c
6.   a
7.   b
8.   a
9.   c.

Chapter 15 (page 000)

1.   c
2.   a
3.   c

4.   b
5.   c
6.   c
7.   b
8.   a
9.   c
10. a
11. b
12. b
13. a
14. b.

Answers to revision papers

Revision Paper 1 (page 179)

1.   b
2.   c
3.   c
4.   c
5.   a
6.   c
7.   a
8.   a
9.   c
10. c
11. b
12. a
13. c
14. c
15. c
16. a
17. b
18. a
19. b
20. c.

Revision Paper 2 (page 183)

1.   a
2.   b
3.   c
4.   b
5.   c
6.   c
7.   a
8.   b
9.   b
10. c

11. b
12. b
13. b
14. b
15. b
16. a
17. b
18. b
19. b
20. b.

Revision Paper 3 (page 185)

1.   a
2.   c
3.   c
4.   c
5.   a
6.   c
7.   a
8.   c
9.   b
10. c
11. b
12. a
13. c
14. b
15. c

16. c
17. a
18. c
19. b
20. b.

Revision Paper 4 (page 187)

1.   a
2.   b
3.   a
4.   b
5.   a
6.   c
7.   c
8.   a
9.   c
10. b
11. a
12. b
13. a
14. c
15. b
16. b
17. b
18. b
19. b
20. c.

Index